RECIPES FOR CLEAN WATER

A Homeowner's Stormwater Survival Guide

Text by William Boudreau

From the Star-Tribune August 1, 1998

Toxic algae suspected at Willmar-area lake

WILLMAR, MINN. — Department of Natural Resources officials are urging people and their pets to stay away from Lake Minnetaga in west central Minnesota while they await water test results.

Fisheries specialist Dave Coahran said a dog and a deer died recently after coming into contact with a blue-green algae. In warm weather, the algae can develop natural toxins that affect the nerves and liver in humans and animals. Symptoms include numbness, dizziness, abdominal pain, vomiting and diarrhea in humans; and weakness, breathing difficulty and convulsions in animals.

The lake is about 8 miles southeast of Willmar.

— Associated Press

Bacteria blamed in Lake Winona fish kill

WINONA, MINN. — Bacteria that occur commonly in southern Minnesota lakes are responsible for a fish kill in Lake Winona last week that left as many as 12,000 crappies dead, the Minnesota Department of Natural Resources (DNR) says.

Flexibacter columnaris occur in background levels in Minnesota lakes, DNR fisheries specialist Dan Dieterman said. He said the bacteria are abundant in nutrient-rich lakes in southern Minnesota where agricultural or urban runoff aid the growth of algae and aquatic vegetation. That can cause oxygen depletions in the water during hot weather, he said, making fish vulnerable.

— Associated Press

Photo Credits: Gary Alan Nelson, Cover (leaves); Minnesota DNR, p.11; Freshwater Foundation, reproduced with permission, p. 20, 21, 24, 25; Chuck Ripper, p. 51; Elizabeth Cedarleaf, p. 55; Donna Krischan, p. 58; Mike Meyer, p. 107

Printed by GraF/X, Minneapolis

Foreword

In August 1995, representatives from neighborhoods surrounding the Minneapolis Chain of Lakes gathered on a site overlooking Lake Calhoun. The beautiful view of the lake, shimmering in the setting sun, was, unfortunately, in sad contrast to the state of the lake's waters. Calhoun, like the other lakes in the Chain, was being polluted by the actions of those of us who claim to love it. The question on our minds was 'what can we do?'

As members of the Environmental Committees of our various Neighborhood Revitalization Programs (NRP), we knew that *stormwater runoff* was the main reason for declining water quality throughout the Chain. So we set out to discover how we, as individuals living in the watershed, could change our time-worn habits — the thoughtless activities that have contributed to the flood of dirty water entering the Chain, dragging down its once-famous water quality. Today, we are proud to publish the results of our work.

The cookbook we've created, with its recipes for cleaning up stormwater, is being presented to the homeowners of Minneapolis - where property values depend on the maintenance of recreational water quality - at minimal cost. And since everybody lives in a watershed and needs to know how to keep from polluting the lake or river downstream, we plan to distribute this information widely.*

Thank you to the members of the Clean Water Task Force, their respective neighborhood associations, and to the staff of the Minneapolis Neighborhood Revitalization Program for their efforts in producing the quality resource you hold in your hands. You've made my job a joy.

Using these recipes will change your diet and your life. Remember, we <u>all</u> live downstream. Please do whatever you can to help us restore the lakes to health and preserve the greatness of our unique natural inheritance.

Jon Carlson, Chair
Clean Water Task Force

*If you move away from the Chain of Lakes, please consider leaving your copy in the house, so the new owners can learn how to prevent water pollution.

RECIPES FOR CLEAN WATER

A Homeowner's Stormwater Survival Guide

Clean Water Task Force

Jon Carlson, Chair
David Shirley, Secretary
William Boudreau
LaDonna Gammell
Jim Daugherty
Robert Kean
Thomas Dicks
Nancy Gross
Cole Fauver
Ronald Fergle
Jeff Shapiro
Mike Elson
Mary Williams
Niel Ritchie
Katy Harding

Minneapolis, 1999

CONTENTS

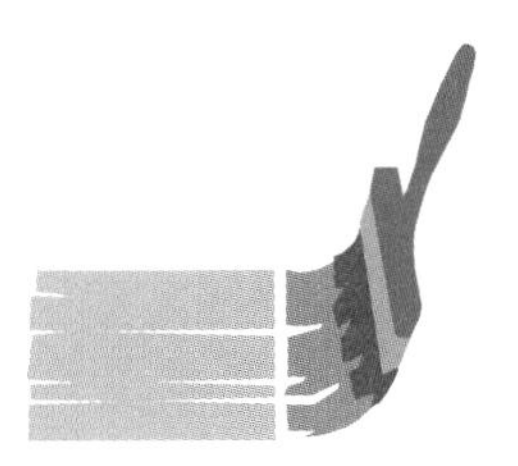

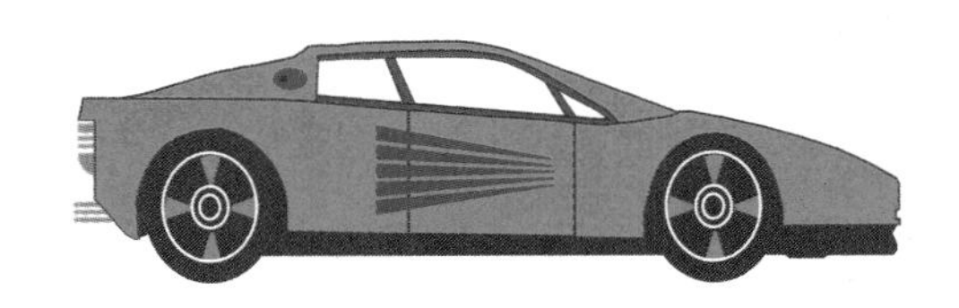

INTRODUCTION

Drip, drip, drip.... With every drop of rain the nation's stormwater crisis deepens, for each perfect crystal bead of water that darts from graying skies or stumbles from curb to street has become a carrier of disease and destruction. No area of the country is spared the plague of bacteria, litter, fertilizer, sediment and over 70,000 man-made poisons that haunt the land.

As humans master the planet, paving over the earth with asphalt, cement, brick and roofing, even Nature's greatest creation - clean water - must give way. Rain and snow visiting our growing cities, are less and less likely to escape unblemished. Denied internal rest in the ground, rain water is forced into the rough society of streets and alleys where it confronts the villains of our modern age: refined oils, heavy metals, yard steroids, specialized chemicals, insect mutagens, pet wastes, loose soils, etc.

Stirred to an angry torrent by concrete curves and inclines, this broad stream of polluted water scours the gutters, sweeping away every disgusting waste and leaving behind a strangely pleasing emptiness. Facing clean streets, we ignore the sad truth that this whole twisted mess is headed for the waters where we swim, fish, drink, and refresh ourselves!

Welcome to the wild, wicked world of **stormwater**. Pretty much unknown when the nation dedicated itself to "swimmable, fishable" waters 25 years ago, nearly half of the nation's lakes and rivers remain threatened by what we officially call 'nonpoint' pollution. Nonpoint pollution, or *runoff*, comes from you and me — the thoughtless by-products of our urban lifestyle.

Sadly, there is no government SWAT team equipped to treat the delinquent flow to our valuable receiving waters. The sources to be turned off reside within our own homes, within our attitudes and customs.

Come with us now as we trace the short but deadly stormwater river, discovering its origins, its destination and impacts and, most importantly, its remedies. Drip, drip, drip...

1
STORMWATER - WHY WE DON'T DO IT IN THE ROAD

One of the few things a majority of Americans agree on is the need to respect and preserve our natural environment. From Walden Pond to Yosemite, John Muir and Chief Joseph to Theodore Roosevelt and Rachel Carson, a singular value of US culture is our reverence toward the awesome landscape that nurtures and sustains us. Despite affluence and occasional indifference, as a nation we have consistently championed a healthy land of clean air and "swimmable and fishable" waters.

When post-war growth threatened lakes, rivers and streams, the country responded legislating in the 1970s against the many points where large volumes of pollutants flowed regularly: factory discharges, contaminated cooling waters, illegal connections and obsolete municipal sewers. It soon became evident that shutting off point sources, while necessary, fell short of the goal.

A second category of pollutants, a mobile wave of hard-to-locate 'non-point' sources, also menaced surface waters. These pollutants (Latin, *pollure* - to dirty or make impure), lurk on farms, city streets, lawns, roofs, and parking lots. Mobilized by rain or snowmelt, they join to form a foul slurry of poisons tumbling to the nearest - you guessed it - lake or river.

It may seem radical to assert that rain is a dangerous compound, like arsenic or dynamite, but

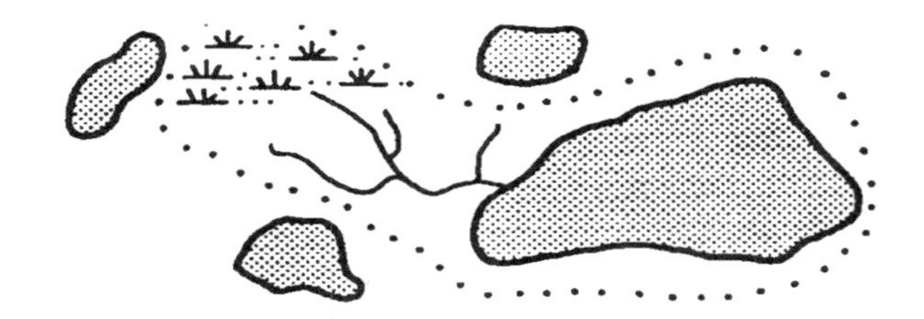

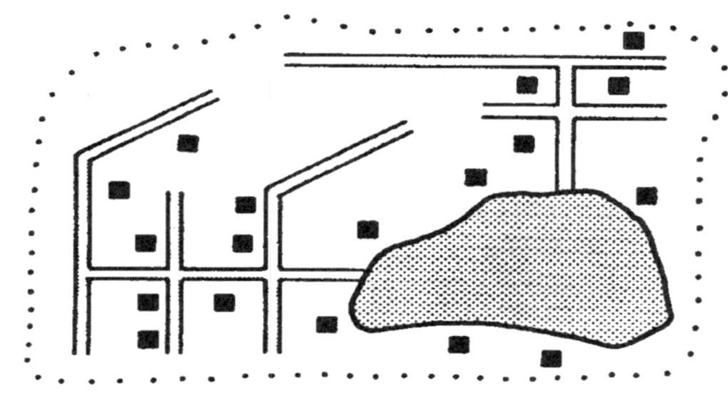

A lake's watershed before and after development.

unfortunately it's true. We've become *homo pollutus* by virtue of our increasing numbers: cities of 100,000-200,000 may have half of their area covered by waterproof surfaces such as roofing, asphalt, or concrete. These materials are designed to be **impervious**, shedding rain and snowmelt to pour downstream.

Acre for acre, a naturally porous habitat such as a meadow can absorb 16 times the tainted rain of a paved parking lot. Once the community grows beyond densities of 1/2 acre residential lots (10-20% imperviousness) however, it faces major environmental damage all the way to loss of whole species.

The stormwater flow is short by river standards, moving only a few miles, but it can be a violent, deadly mixture. The Environmental Protection Agency (EPA) calculates that nonpoint runoff pollution is the most complex pollutant known, composed of 137 chronic (long-term) and toxic (short-term) killers.

Sometimes we forget that our remarkable network of all-weather streets were designed to transport more than seated humanoids. We don't see that the crown of the street has less to do with the regal regard accorded the automoble, than with the need to avoid puddling. We can't feel the separate universe pulsing beneath our feet. We don't understand that our streets, built as a liquid evacuation system, have morphed into a 'pollutant resource pool.'

Stormwater pollutants often arrive through the odd cooperation of natural forces and unwise cultural practices. The world's skies for instance, imprison a variety of micron-sized (one-millionth of an inch) pollutants blown off the planet's surface on hot, upward gusts. Spread over the globe by jet-force winds, these dangerous hitchhikers mill about a mile thick waiting to make a break. As the heavens are periodically wrung out, they seize their chance, smuggling themselves aboard a droplet of rain or snow to sew chaos on the surface. Among the more common air-terrors:

Mercury - from garbage incinerators; coal and wood burned for power, heat and cooking;
PCBs (polychlorinatedbiphenyls) - from synthetic oils employed in electrical transformers, cutting oils, and carbonless paper;
Dioxin - from incinerators and chlorine use in papermaking;
Soil particles - from agricultural areas;
Pesticides - from across the globe;
Nutrients - nitrogen and phosphorus from fertilizers and pollen; and,
Sulfuric acid - from smelting and power generation.

All are frequent, destructive, stowaways.

Of course, rain falling on grass or shrubbery slides into the earth carrying with it the worst airborne critters and rendering them more or less harmless (small quantities of *nutrients* are actually beneficial to plants).

But a raindrop smashing onto bare ground may impact so violently that soils fuse into a solid surface. Repeated landings, unable to be absorbed, begin to puddle. It's here that the inmates from the sky merge with the gang from the ground, blending into a ghostly tide rolling across the land.

A construction project, one of thousands on-going during warmer months, is the first target. This one includes a new lawn, a garage, patio and a concrete driveway.

The garage excavation has created 'extra' soil which is dumped in the street with a crude

message "Free Dirt" staked to it. A new, more impervious, concrete drive replaces the existing porous crushed rock.

The lawn too, has been torn up and piled haphazardly. As rain intensifies pounding the site, the old sod is stirred to a chocolaty froth, bleeding a thin black ribbon of soil down the new concrete spillway.

Nearby, two long fluorescent lightbulbs and a discarded thermostat are propped against the garage. These items contain mercury and, by law, cannot be incinerated or landfilled. But if they're just set out next to the alley and something should happen to them ...

Meanwhile, rainwater shoots out of

downspouts aimed at the impervious apron of fresh concrete with the force of a water-cannon. The fluorescents stagger and crash spewing gruesome mercury and phosphor powder into the fast-moving stream.

On the flanks, progress is slowed by turf culture. But this is no longer a diverse native community of deep-rooted, climate-specific grasses, forbs and sedges reknowned for their vigor under near-drought conditions. Nor is the topsoil airy and crumbly, rich in minerals and decomposing plant material (*humus)*.

Instead the vegetation is a fast-growing variety of grass known as Kentucky blue, *Poa pratensis,* a high-need child neither blue nor Kentuckian. And the original soil has also been removed, replaced by a moisture barrier of compressed clay tilted at a 2 degree angle *toward the street...*

Rainfall can be absorbed by this 'lawn carpet' for only a short time before it is saturated and water begins to run off down the slope. Trickles merge into rivulets, rivulets give way to streams, streams swell into rivers, transforming enormous volume into powerful flow energy.

With increasing strength, the stormwater surge strips the sod of any not-yet-absorbed *phosphorus* and *nitrogen* breaking it down and bearing it away.

Pesticide residues, complex, constant chemical killers such as aldrin, dieldrin, octochlor, etc. are next to be purged from the artificial landscape, helping to form a deadly tide creeping toward the sidewalk.

A cargo of eroded soil from bare areas of

RECIPE FOR CHAIN OF LAKES DISASTER

50% impervious surface
60 lb. heavy metals
zillions of bacteria
600 lb. ethylene glycol

30,000 lb. sand
250,000 rough fish
1,000 kg. phosphorus
2.5 million visitors

Mix ingredients in 350,000,000 gallon bowl. Add huge dose of fertilizer from bottom muck and let stand one year. Yield: lethal fall-out, shifting, unstable sediment, officially "threatened" swimming, more weeds and lost water clarity.

the lawn where traffic or the mower has cut deeply, clouds the water with fine particles called *silt*.

On the concrete walk, the pace quickens, joining a steady flow off the driveway surging with a slug of *grass clippings* and a film of *detergents* used to clean the family car. Several floating *cigarette butts* mark the river's progress as it runs off the walk and pours onto the bare, compressed ground of the boulevard.

Here, the rush of rain encounters a fresh pile of dog *excrement*. At first the mound is resistant but the current gradually undermines it carrying away the various *bacteria* and *viruses* and *nutrients* that make an uneasy home in animal intestines.

Briefly, the runoff pools behind the lip of the curb. As it ponds, it pauses. Below, the pulverized *litter* of the street stirs to life.

Grease, *motor oil* and other 'auto-matic' secretions like *ethylene glycol* (antifreeze) have spent the past week congealing over a vast herd of dirt and grit. With the first flush of precipitation, this gooey mixture springs to life. Soon the familiar petroleum rainbow snakes its way along the gutter ready to strike and poison at the astounding rate of 1,000,000 parts fresh water to 1 part oil. Nor is this the last word from Detroit.

Further pollution occurs as flakes of the heavy metal group — *lead, zinc, copper, chromium, cadmium and nickel* — from the body, exhaust and tires, lose their grip in the auto industry and fall to the street. Fierce pressure from thousands of daily trips bond the specks with runover dirt (*fines)* forming a sullen army awaiting one last downhill power drive. Our late-afternoon thunder shower taps them to join the stormwater riverdance.

The skies unfurl now in sheets of water. Global atmospheric pollution piles onto local pollutants already on the ground constructing a massive flood of filthy water.

Near the end of the block, a thick mat of leaves ripe with *nutrients* forms a dam along the gutter. Smooth concrete channels supply speed and muscle to stormwater until, like a wire brush, it smashes the leaf dam, scouring the curb. Up ahead, the stormwater cataract roars, angrily hurling itself against the steel teeth of the corner sewer grate.

Great chunks of necessary habitat are swallowed in the downstream current. Sliding away too, toward a network of waiting storm drains, strategically engineered and placed, are the last remnants of a self-sustaining, self-replicating ecosystem.

CHAPTER 1

Suddenly, the river disappears, detouring downward several feet into the dank world of rotting, fetid underground pipes. It races now, the area alloted for its passage narrowing even as the volume of untreated stormwater increases. In just a few moments, this curtain of impurity leaps from 4-foot wide tubes onto the backs of aquatic life in lake and stream.

* * *

Stream riffles, tiny nooks where insects find shelter, disappear under layers of silt. Deep, oxygen-rich pools fill with contaminated *fines*. The wetted stream border, with its cooling overhead cover required by frogs, salamanders, turtles and other wildlife for both food and reproduction, is destroyed by short-lived stormwater floods pushing stream channels 8" above normal flow levels. Water temperatures rise by as much as 10 degrees.

As the tons of leaf and lawn litter begin to fall apart at the end of the stormwater pipe, they absorb large amounts of oxygen from the water. Many invertebrate members of the aquatic food chain - immature nymphs of the stone fly, mayfly and caddis fly - cannot withstand the loss of refuge and the warmer, oxygen-poor water. Unable to reproduce, their extinction seals the doom of sensitive fish species like trout and sculpin. New species, warm-water, pollution-tolerant ones, such as tubeworms, biting flies and snails, take over.

* * *

Lake effects are just as devastating. Nutrients phosphorus and nitrogen arrive in enormous quantities — increases of 100,000 tons per sq. mi. per year accompany the change from forest to small residential lots, big-box merchandisers and clover-leafs. Unwise development means many urban lakes receive <u>annual</u> phosphorus loadings equal to all the background and atmospheric phosphorus deposited in their previous 10-12,000 years of pre-development history. And, unlike streams which can 'cleanse' themselves of flowing pollutants, lakes often don't have outlets. Pollutants may remain in sediments for years, becoming part of the in-lake nutrient cycle.

The barrage of stormwater runoff pollutants produces dramatic effects in *mesotrophic* lakes, or those of "middle-fertility" — generally regarded as the 'best' recreational waters. These lakes, with swimming beaches, fishing docks, stands of bullrush and water-lilly, wash away sweat, toil, and cool the brow and spirit of the masses. They put the heavens within reach and link the human population with the universe at large with sweet, refreshing caress.

Unlike deep, cold, little-nourished *oligotrophic* lakes, 'middle' lakes already circulate a fair amount of nutrients. These nutrients support a *littoral*, or shoreline environment rooted, literally, in plants. *Macrophytes*, or rooted plants, are important to lake ecology. Their extensive system of shoots and tubers binds lake muck keeping water clear - a necessary condition for the survival of prey species, like minnows. Their broad leaves slow wave action stabilizing the shallow-water world. Rooted plants also take up phosphorus that might otherwise spur growth of tiny, free-floating plants called *algae.*

Algae grow and reproduce by converting sunlight and available phosphorus and nitrogen into fuel for cell construction and other bodily processes without the need for soil anchoring. Among the most successful of all plant species, algae boast a family tree with thousands of branches and roots stretching back at least a billion years. Numerous types of algae are present in virtually all waters and are a necessary first step in a healthy food chain.

In lakes of stable, middle fertility, algae are kept in check by competition with rooted vegetation for limited amounts of phosphorus, and also through active grazing by small, barely visible water animals known as *zooplankton.* Zooplankton provide the initial food source for what become a lake's largest finned predators.

Before the arrival of Europeans, most lakes in North America were in this state of "trophic equilibrium" with little change in the physical relationships between algae, its competitors and predators.

But the presence of humans at lakeside has had pronounced destructive effects on water quality documented as far back as Roman resort construction during the reign of Julius Caesar. Depending on the character of the watershed, harmful effects could result from no more than the replacement of evergreen trees with deciduous trees and their greater amounts of leaf litter.

This ill-advised placement of soil awaits a "timely" rain before it joins the lake.

Lakes assaulted by enormous amounts of polluted runoff frequently become unbalanced, rapidly declining in water quality and limiting their recreational uses. This speed-up in lake fertility due to nutrient enrichment is known as eutrophication (Greek for "well-nourished").

Eutrophication (pronounced *u-tro-fi-ka-shun*) is a well-understood and predictable process. Lake scientists, or *limnologists*, simply measure the volume of the lake, the amount of nutrients going into it, and its current water transparency, to arrive at a set of numbers that describes the rate of eutrophication for that particular body of water.

When eutrophication affects use of the lake by fish, wildlife and humans, a use problem, or 'impairment' exists. Stormwater is the chief eutrophier of lakes today, traced to virtually all impairments.

Stormwater dumping into shoreline areas heaps polluted material hundreds of feet into the lake. The shifting, sand-like fines are too unstable to root aquatic plants. Excess phosphorus is taken up by algae. Algae suddenly "bloom" into large surface scabs which can blanket shallow areas, warming the lake and preventing sunlight from reaching the bottom. Loss of sunlight further interferes with the germination of rooted plants.

The silty shallows are easily disturbed by wave action, particles suspending in the water column where they create constant turbidity.

Gamefish like pike, bass and walleye are sight hunters, they avoid murky water where their excellent eyesight is of little advantage.

Algal hordes gorge on phosphorus until it's gone, then they die, sucking the last bit of oxygen from the water. With large areas of the lake unavailable, fish don't grow as fast or as large as they could.

Clouds of silt now slide toward the only stretch of spawning gravel left on the lake bed. Dirt rains down on clusters of walleye eggs, smothering them.

The stage is set for a major shift in lake inhabitants. Oxygen-loving, cool-water species find less and less of the lake liveable. Bullheads and carp, tolerant of higher temperatures and much less oxygen, find their range expanded. Carp are especially obnoxious in a recreational setting, ripping up large tracts of rooted vegetation in search of aquatic worms. Carp are nutrient pumps, their short intestines producing an ammonia-like fluid so rich in phosphorus, their wastes may nearly equal nutrients from runoff sources.

While most high-quality lakes have neutral or slightly acidic water chemistry (pH at or below 7), carp feces are extremely alkaline, pushing waters in which they are plentiful perilously toward the base end of the scale. Extremes in pH are deadly, and *green algae*, a vital food link between aquatic plants and animals, begin to disappear as pH heads upward. Abundant phosphorus is now taken up by *blue-green algae*, a variety that thrives in overly-fertile, high pH waters.

Blue-greens are more resistant than greens to grazing by zooplankton, particularly the large-bodied *daphnia*, or water flea. Daphnia consume immense amounts of algae but only if there is enough oxygen at lower depths so they can escape being eaten themselves by small sunfish, perch and bluegills. Low oxygen, high pH and large blue-green colonies disrupt zooplankton feeding cycles necessary to clear algal blooms in mesotrophic environments.

With the trophic balance tipped in their

favor, the opportunistic blue-greens spread their hair-like filaments throughout the shoreline, reaching toxic concentrations in stagnant waters. A golden retriever, a family pet, frolics in the lake despite warnings. After running down several sticks in the mucky shallows, the dog drinks deeply beside the slowly rotating skeletons of a snowy egret and a great blue heron. The retriever bounds away following a whistle but soon feels dizzy and lies down. He never gets up.

Poisons attached to the silt - cadmium, zinc, lead, etc. - kill off small insects - vital links in the food web upon which desirable game species depend. Other poisons - mercury, PCBs and dioxin - are passed up the food chain to humans. Mercury concentrations now are 40 times greater than 100 years ago.

Once in a lake, mercury is converted to *methylmercury* by bacteria. Fish absorb methylmercury as it passes over their gills, or by predation on smaller fish and aquatic animals. No method exists to reduce mercury in fish. Humans who eat contaminated fish may develop health problems over time. Early signs of toxic mercury levels include a loss of coordination, and a tingling sensation in the finger tips. Children and unborn fetuses face the greatest risk. A growing number of lakes tested report dangerously high levels for mercury. Fish consumption advisories for children and women of child-bearing age have been issued for these lakes.

Aquatic plants absorb PCBs and dioxin from stormwater-tainted sediments. Eventually, these toxics make their way to large fish and then to humans. Trimming away fatty fish tissue will lessen the risk of cancer in adults. Again, children are at greatest risk, facing developmental problems at threshold levels. Fish-eating birds such as cormorants face severe damage -- bent beaks, immature wings, lack of sexual maturity -- all from poisonous runoff.

No resident of the lake goes unaffected by the equivalent of untreated sewage oozing into their environment. While you may not remember any illnesses from those childhood popsicle regattas in the gutter, your offspring are playing at the beach near a new 96" outfall pipe. Out of that cylinder tumble a variety of violent viruses: *Hepatitis A* ; *Cryptosporidium*; and a battery of bad bacteria: *salmonella*, *staphylococcus* (staph), *campylobacteriosis*, etc.

In the 48 hours between being coughed out of the sewer and their demise at the hands of burning solar rays, these invisible predators wait to be swallowed. Children make excellent multiplication chambers. They boast numerous entry points - scratches, mosquito bites - and they swallow a lot more water than adults. After a morning in the water, your eyes leak a sluggish yellow fluid and your nose feels stuffy; the kids look pale and are moving slowly, complaining of upset stomachs. You pack the car and head for home but on the stairs your daughter falters and you carry her and the secret colony building inside her to the couch. She falls silent a moment then sits bolt upright. She coughs and a stream of green vomit flies across the room.

Back at the beach, a scientist fills a beaker with beach water to determine the amount of *fecal coliform* present. Too high a count indicates the likely presence of disease pathogens found in the stomachs of various mammals. The allowable standard of 200 parts fecal matter per 100 millilitres (1/2 cup) of water is exceeded a hundredfold: 20,000 particles!

The scientist performs another simple test: she sheds shoes and socks wading into the water to about 4'. She looks down in the water searching for her toes. The lake is dark green, her feet unseen. She leaves the beach for her truck, retrieving a sign employed several thousand times each year in the U.S.

Beachgoers congregate, then slowly move out of the water and fold their towels. A sign is pounded into the shoreline scum: Warning! Contaminated Water. Beach Closed by Order of the Department of Health.

2
STORMWATER POLLUTION A CASE STUDY

The Minneapolis Chain of Lakes

The Minneapolis Chain of Lakes receives significant stormwater runoff — about 40% of its annual volume — and is typical of the 30 designated receiving waters within the city. The Chain consists of four small recreational lakes — Cedar (190 acres), Isles (102 acres), Calhoun (421 acres) and Harriet (353 acres).

The first three lakes are linked by navigable channels and comprise the upper chain at 141' above sea level. Lake Harriet, the only city lake classified as non-eutrophic, lies 1/4 mile downstream of the upper chain and is four feet lower in elevation.

These "equisitely beautiful sheets of water" as early settler Henry Hunt Snelling (of the Fort Snellings) described them (1888), are ice-block lakes left by the most recent glaciation 12,000 years ago. Lying above the water table (only Harriet receives significant amounts of seepage from groundwater), the upland environment surrounding and draining into the Chain is extremely important as a source of lake recharge and potential pollution.

CHAPTER 2

The 1850's map (right) shows the extent of wetlands protecting the Chain (errors may exist due to surveying techniques at that time). Compare with the modern Chain (next page) which lacks protective marshlands. (Green areas are park and/or cemetery.)

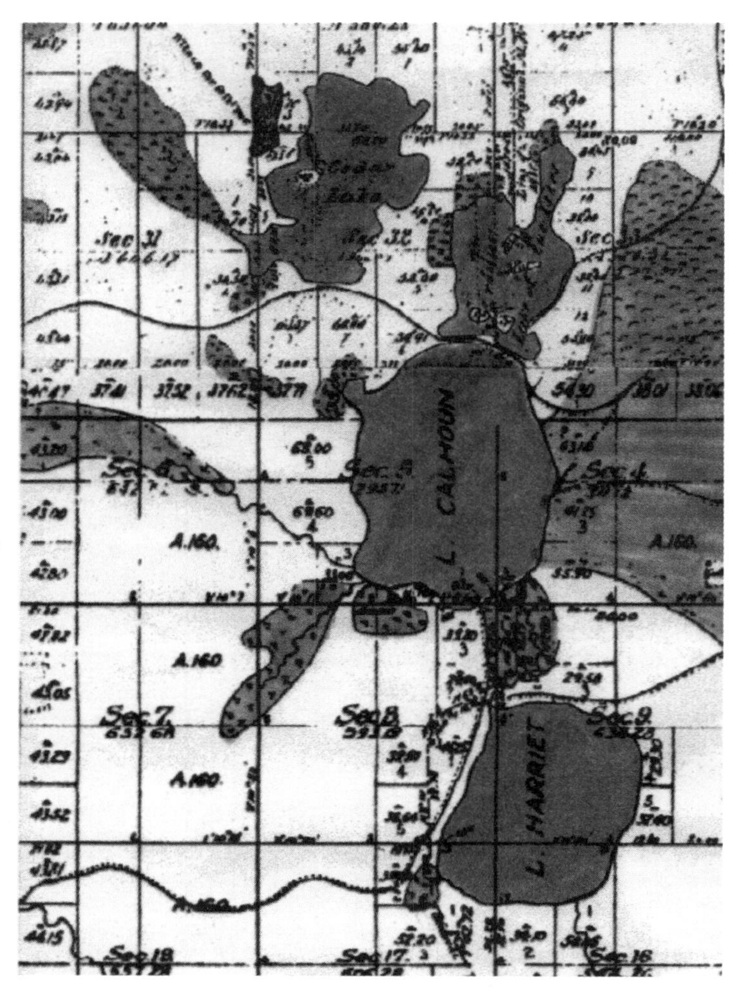

The earliest recorded human settlement of the Chain dates to Chief Cloudman, a Dakota warrior, who led a small farming village on the shores of "Mde MA-KA-SKA" - Lake of the White Earth (today's 'Calhoun') in the 1830's. When white settlers arrived, Cloudman bid them join the community, offering them a prime spot along the east bluff so they could view the remarkable bird life.

Europeans found the lakes to be "of the most Eden-like loveliness...clear and transparent...in the deepest part." Sediment cores taken from the lake bottoms showed that both Cedar and Calhoun tended to oligotrophic, or low-fertility lakes before development, with water clear to a depth of 10' to 15'.

The youthful city which grew up around them celebrated its freshwater bounty in the municipal slogan 'City of Lakes.' But the young metropolis soon boomed to a population of 550,000 (1950). Extensive settlement of southwest Minneapolis produced drastic changes.

Isles, the second and smallest lake, was little more than a marsh before two decades (1889-1911) of dredging deepened its basin and connected it by channel to Cedar and Calhoun. The excavations cost Cedar 9' of depth and many acres of vegetated shoreland. Natural wetlands protecting both Cedar and Calhoun were filled in, trees cut down, sod removed, and streams straightened.

Tall native grasses gave way to short-

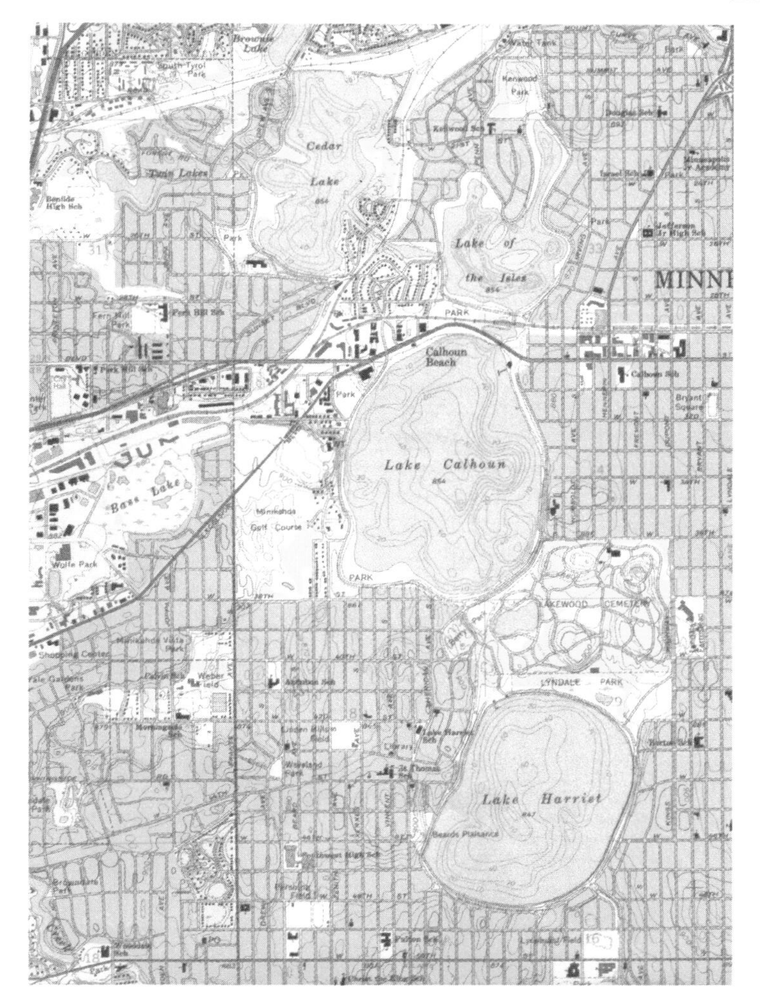

leafed turf grasses drugged with phosphorus and nitrogen from unnecessary lawn fertilizers. More lawns meant more grass clippings in the street. Raking leaves into gutters became the disposal method of choice following the leaf-burning ban of the Environmental Decade.

By 1970, half of the surrounding 6,968 acre watershed draining into the lakes was developed to 50% impervious surface. Over 1 billion gallons of stormwater pours into the Chain each year. The lakes need this water to maintain themselves, but they've few defenses against the massive undertow of pollution that accompanies it: urban runoff contributes half the water but 95% of the 3.5 tons of phosphorus that slumps into the Chain annually. Phosphorus grows algae at a ratio of 100 to 1, yielding 350 tons of sticky green stuff each year from runoff alone.

Each spring according to Public Works estimates, 15,000 tons of sand put down by city crews over winter, is "lost" to the Chain and the Mississippi River.

Flooded basements in low-lying areas of the city were frequent because the land had lost the capacity to hold normal rainfall. The routine commingling of millions of gallons of raw sewage with runoff waters led to lengthy and costly legal battles with downstream neighbors. Even the names of the lakes had changed, dropping their melodious Indian flavors for, in the case of 'Calhoun,' the name of the man who started the American Civil War.

Biological communities stable for eons suddenly found their ecosystems under attack by all sort of 'eutrophiers.' Water quality, our nickname for the Chain's ability to cleanse, console, entertain and inspire, degraded at fast-forward speed.

The first major study of the plight of Minneapolis lakes, completed in 1969 by Hickok and Assoc., hydrologic engineers, concluded that the Chain had experienced an awful turn-around, shifting to "eutrophic," an environment which favored plant over animal life!

Hickok took a unique approach to analyzing city lakes, surveying bottom life as an index of water quality. It works like this: bottom fauna are sensitive to changes in the environment, directly reflecting pollutional effects or stage of eutrophication. Clean, non-polluted water is inhabited in its depths by a wide variety of biological life but only a few specimens of each type.

Unaltered photo of runoff to west Calhoun shoreline.

Polluted water on the other hand, has only a few different types of organism, but many more specimens of each type.

Clean environments foster competition and offer a variety of niches to aquatic life; dirty environments are tolerated by far fewer species but these 'successful' types proliferate. A non-polluted area is one in which the clean-water associated organisms comprise more than 50% of the total kinds of organisms present in a given community.

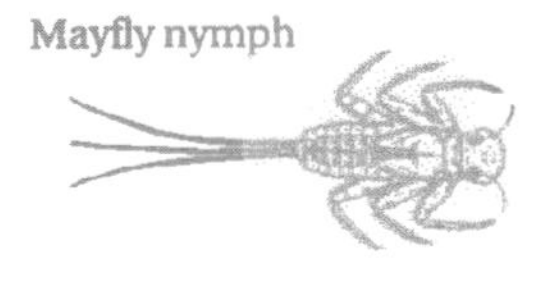

Pollution sensitive:
mayfly nymphs
caddis fly larvae
gilled snails
hydra

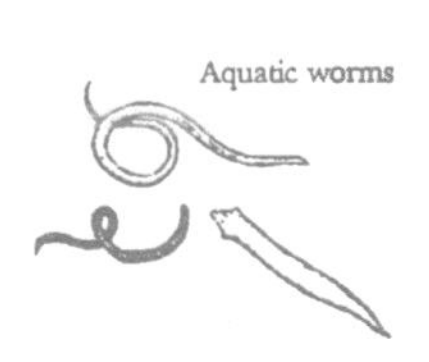

Pollution tolerant:
scuds
sludgeworms
bloodworms
damselflies

The majority of the organisms found in all lakes in the survey were considered pollution tolerant. Pollution sensitive creatures were discovered in Cedar and Calhoun but they comprised less than 20% of those animals present.

In addition, all lakes had sufficient phophorus at the start of the growing season to spawn "abundant" algae blooms of the nuisance type. Plus, Calhoun's dissolved oxygen in March wasn't sufficient to support fish below a depth of 25 feet.

Hickok's work came at the height of national environmental concern over the growing surface water crisis in America. Industrial poisons were routinely poured into rivers already filled with raw sewage; Cleveland's Cuyahoga River became the poster child of the deplorable situation when it caught fire in 1969. By 1972 environmental sentiment culminated in the Clean Water Act (CWA) - federal legislation to police the unrestricted dumping of industrial, municipal and residential wastes into the country's precious water resources.

To implement Congressional aims, the National Pollutant Discharge Elimination System (NPDES) required the states to eliminate all pollutant discharges to local receiving waters. Initially, government resources targeted point sources - factories, over-worked treatment facilities, illegal connections, etc. While progress was made, it became evident that 'non-point' sources - you and me and the millions who littered into the streets - were also a major obstacle to CWA aims of "fishable, swimmable" waters.

Prof. Joseph Shapiro of the University of Minnesota's Limnological Research Center, arrived at that conclusion after a comprehensive analysis of the Chain (1971-72). Using historical

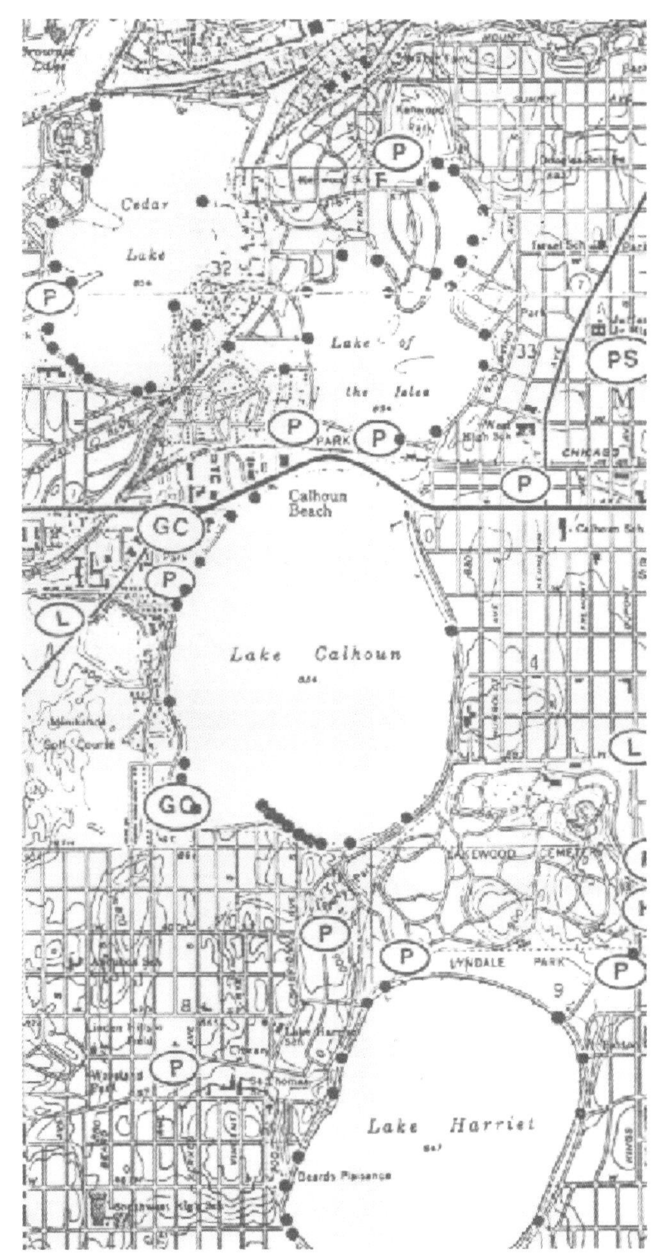

Some 75 outfall locations dump into the Chain; nearly 300 empty untreated runoff directly into the Mississippi river.

records, and core samples from lake sediments, Shapiro concluded that all the lakes had declined "significantly" from their pre-development days. Dissolved oxygen, nitrogen, chlorophyll, phytoplankton (algae), pH values and rates of photosynthesis in all the lakes were typical of highly

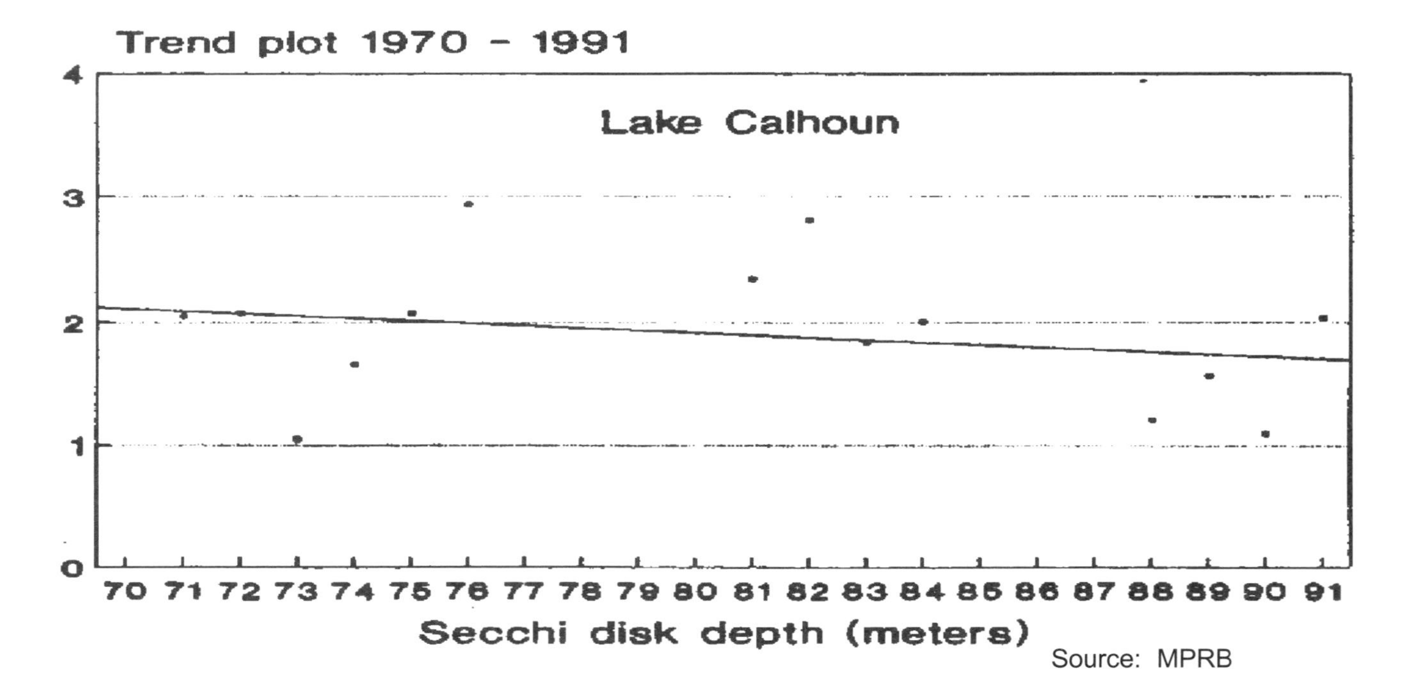

Source: MPRB

eutrophic, or very fertile, lakes.

Shapiro also outlined reduced habitat for gamefish, and phosphorus loadings three to seventeen times the amounts necessary to convert oligotrophic to eutrophic lakes. Shapiro noted that in the first half of 1971 alone, phosphorus in runoff to Calhoun amounted to 1.9 metric tons, or 3.5 times the threshold loading distinguishing a eutrophic Calhoun from an oligotrophic one. Historical data, said Shapiro, indicated a 10- to 100- fold increase in algal biomass in the past 50 years, and a 17-fold increase in sediment accumulation rate in the past 100 years.

(Interestingly, Shapiro conducted tests on nearly 100 lawns in the Lake Harriet watershed. All samples revealed sufficient phosphorus already present to promote adequate lawn

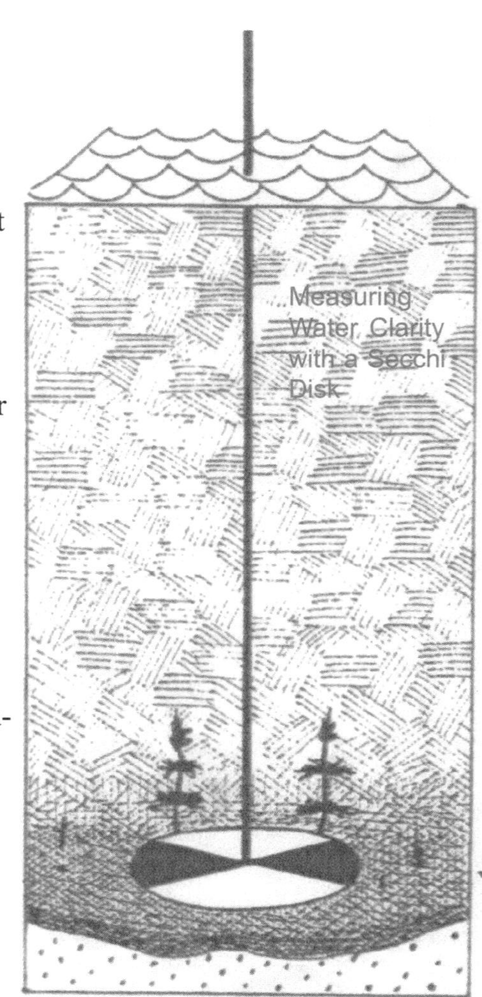

Lake transparency is measured by means of a known length of rope connected to a plate-sized disk, a Secchi disk, lowered into the water until it disappears. The depth at which it is no longer visible is that lake's Secchi disk (SD) reading.

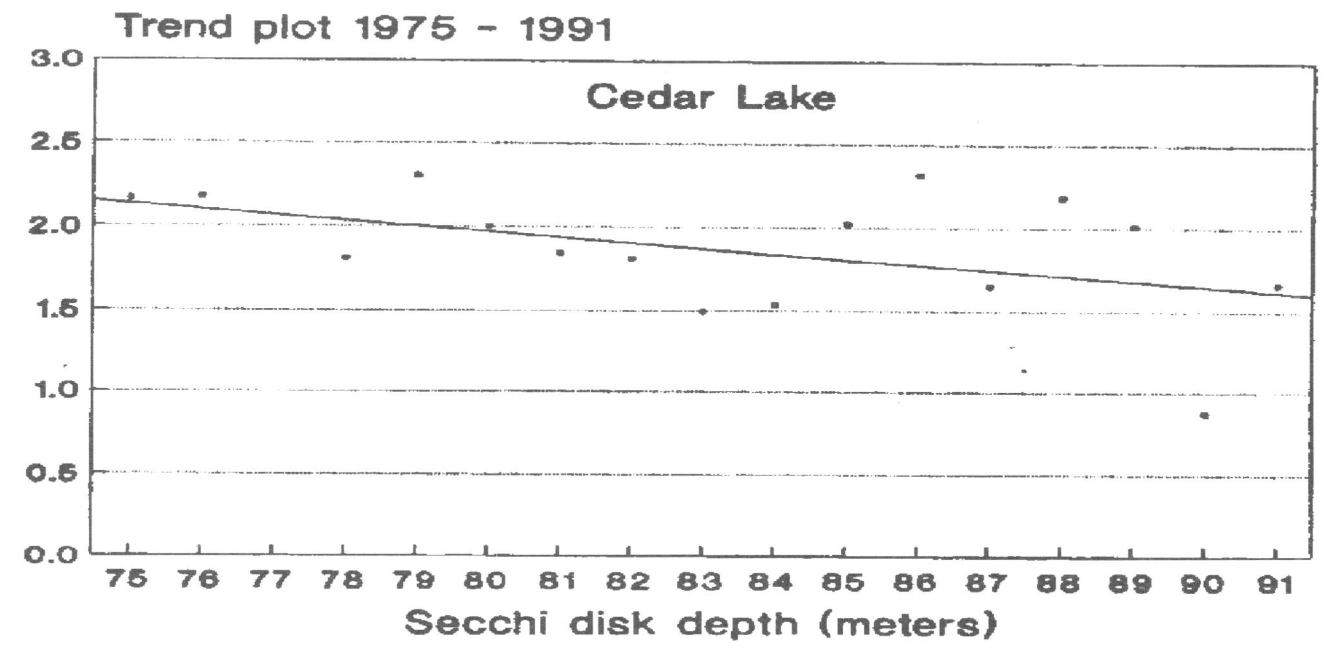

Source: MPRB

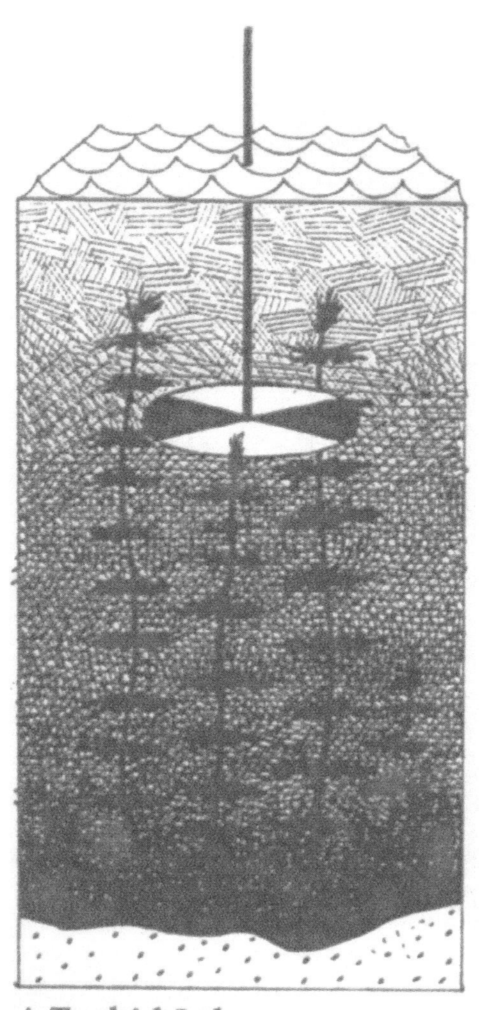

A Turbid Lake

growth. Harriet could have been saved 200 lbs. of unnecessary nutrients from these lawns alone.)

Shapiro's investigation concluded that the chief cause of increased lake productivity and their aesthetic deterioration "is the channelling of storm drainage with its high concentrations of algal nutrients to the lakes beginning in the late 1920's." Shapiro outlined a series of acts he believed could produce a return to 1927 conditions.

As a follow-up, a 1974 Minneapolis Lakes Citizens' Advisory Committee on Lake Water Quality was convened by joint city council and Minneapolis Park & Recreation Board resolution. The Committee specifically advocated the restoration of lake transparency to "1927 levels."

The 1927 readings are important not only as the earliest existing records for the Chain, but also because they depict a Chain strikingly different from that on view today. Compared to 1927 data, all the Chain have experienced profound change, moving down one trophic category. This kind of deterioration within two generations can only be described as catastrophic.

Shapiro, a Committee advisor, promised to improve the Chain "very close" to 1927 clarity in 6-10 years with "a complete management program." In response, the Committee incorporated some Shapiro recommendations and added their own energetic schemes.

Restoration, of course, means returning the complete ecosystem — structure and function — to the original self-regulating resource which existed prior to disturbance. It does not mean tinkering, trade-offs or simply reducing the supply of offending stormwater. Common sense however, says that restorations cannot be perfect. There is little chance

Lake Categories: Eutrophic 1.5 - 6.5 Mesotrophic 6.5 - 12 Oligotophic 12 +	**1927**	**1958**	**1972**	**1992**
Cedar	10	7.5	6	5.4
Isles	8.5	9	2.8	2.6
Calhoun	11	15	7	5.9
Harriet	13.5	6.5	9	8.3

Given here are Chain water clarity readings gathered over the last 70 years. Clarity parameters for lake types are on the left. All readings are in feet.

of recapturing original soil and water chemistry, just as there is no way of reintroducing plants and animals except in approximate relationships.

As long as stormwater remains dirty, any "restoration" would be doomed. And unless Shapiro's "complete management" policy were followed through, stormwater quality couldn't improve significantly. Unfortunately, the expensive and technical nature of the subject did not easily translate into a political mandate. It's known as the 'stormwater stalemate.'

Three small-scale lake improvement experiments were conducted in response to Advisory Committee requests. In 1978, Hickok & Assoc. used *vacuum sweeping* in the Lake Harriet watershed (843 acres) to collect "fines" not picked up in the biannual brush street sweeping. Fines and street debris were identified as the primary source of nutrient enrichment of the Chain.

Hickok also conducted a *first-flush diversion* of stormwater in the Lake of the Isles watershed. Diversion of the first hour of runoff - the period in which most pollutants are mobilized by the stormwater river and dumped into receiving waters - would allow initial flows and flows from minor rain events to enter sanitary sewer pipes for treatment. Subsequent, cleaner flow, would continue to drain, untreated, to the lakes.

'74 LOW-CAL COMMITTEE CONSUME

10 miles "first flush" diversion
33 silt deltas dredged
100 hours public education
5 aggressive municipal ordinances
5 'camels'
50 cu. yds. rock rip-rap

Divert stormwater to sanitary sewer treatment; add dredging, shoreland erosion control, vacuum cleaning vehicles ('camels'); mix in rules and information. Yield: Lake restoration.

While regular sweeping removed 418,000 lbs. of organic material (leaves) from the Harriet watershed in 1980, vacuum sweeping collected and removed an additional 71,000 lbs. of which

250 lbs. was phosphorus. The phosphorus removed by vacuum sweeping was 17% of the total entering Harriet during the year and produced a decrease in Lake Harriet total phosphorus — *unlike the upper chain.*

Diversion of low-flow runoff at Isles also resulted in a reduction of phosphorus to the Chain — 15% concluded Hickok. However, a combination of selective vacuum sweeping and diversion, produced a 27% reduction in the total annual loading to the Chain, or an estimated 2-foot increase in lake transparency throughout the Chain! The cost for these measures was estimated at $4 million over 10 years.

The third experiment, manipulating the numbers and types of fish and tiny plankton, was supposed to make *daphnia* more efficient grazers of algae, thus leading to a cleaner lake without expensive measures to reduce nutrients. The idea is simple enough: introduce large fish that eat the small fish that eat the daphnia that eat the algae that eat the phosphorus that live in the house that Nature built.

Shapiro, an expert in the technique known as *biomanipulation*, proved in Wirth Lake (1977), that removal of plankton-eating fish and nutrient-recycling bottom feeders could oligotrophy an environment by removing predators and enlarging the oxygen-rich zone required by the algae-grazing water flea. While the supply of phosphorus from runoff remained the same, Wirth Lake water transparency improved from eutrophic (5') to oligotrophic (14') due to the selection of the right kind of creatures.

Shapiro hoped to duplicate results in Lake Harriet. Department of Natural Resources fisheries personnel however, weren't willing to allow poisoning the entire lake or, alternatively, prohibiting the taking of top predators. Instead,

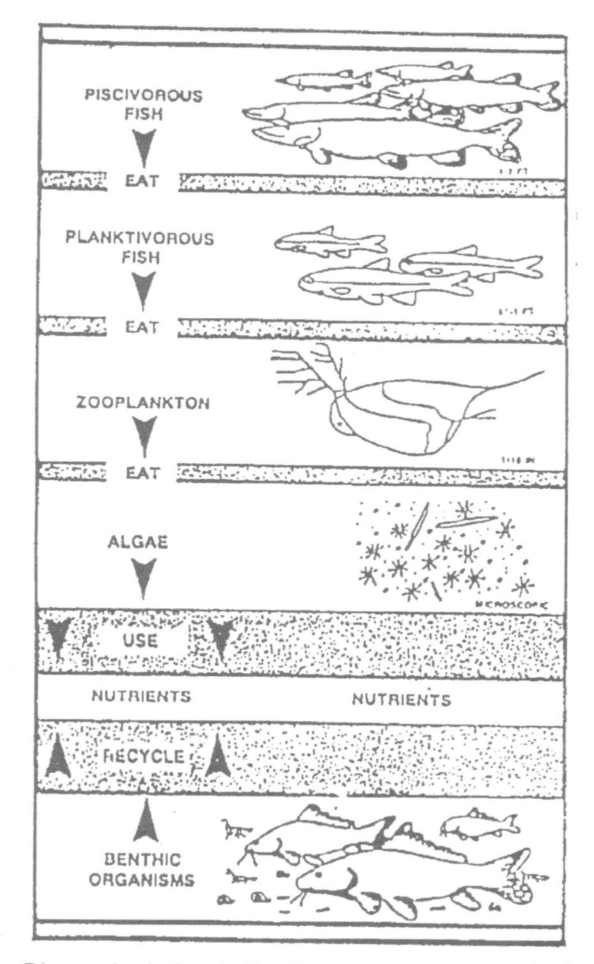

Biomanipulation is the human management of species to produce a stable, balanced, biological community in a clean lake environment.

muskellunge, a larger, hungrier relative of northern pike, were stocked while pike spawning was eliminated.

The effects on water quality were inconclusive. Harriet retains decent overall water character and, although its clarity shifts, it's difficult to determine without continuous testing whether this is a result of varying amounts of polluted runoff or in-lake biological changes.

One of those in-lake changes was the arrival of Eurasian water milfoil, a fast-growing macrophyte which has "clarified" the Chain in recent years by absorbing available nutrients. That's the upside. The downside is dense mats of vegetation spreading at the rate of 2 inches per day across the surface, the disappearance of native species, and severe oxygen losses threatening aquatic life when the plant undergoes fall die-off.

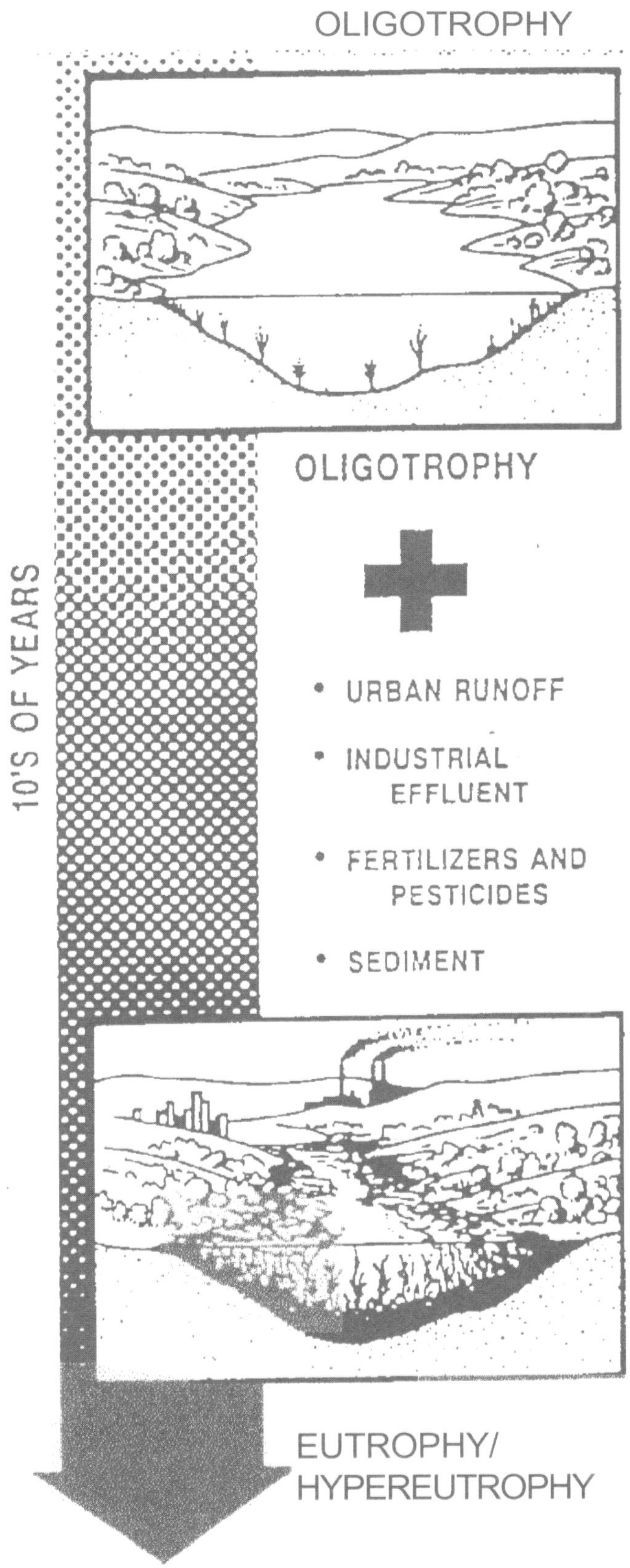

Despite some promise, these small-scale interventions fell short of a necessary comprehensive strategy. The problem did not go away and, in 1978, the Environmental Protection Agency (EPA) commissioned a Nationwide Urban Runoff Program (NURP) to assess more precisely the threat of 'nonpoint' surface water pollution. NURP identified **stormwater** as the culprit in the impairments that today imperil <u>half</u> the nation's lakes and rivers.

Acting on the NURP findings, the EPA sought a new mandate in 1987 amendments to the CWA of 1972 that required major municipalities to secure permits for their stormwater discharges. The permit process required Minneapolis to inventory all stormwater discharges to receiving waters analyzing samples for 77 "priority pollutants" NURP identified as frequent passengers in stormwater. Additionally, the City of Lakes was again required to set up a citizen apparatus to propose a program of water quality improvement.

The scientific portion of the permit completed in 1991 reiterated earlier findings: 1) the range of concentration of all pollutants - toxic metals, pesticides, hydrocarbons, poisonous organics, oxygen-demanding nutrients, fecal bacteria, mercury and sediments - is consistent with nationwide NURP findings; 2) phosphorus and nitrogent input into the Chain from stormwater was high enough to cause water quality degradation; 3) trophic, or fertility levels, in Calhoun and Cedar show these lakes are well into the eutrophic range approaching levels which will not support swimming, sailboarding and other contact recreational uses; 4) phosphorus loadings to Lake Calhoun exceed permissible

TOP: Elevated sites & steep slopes require extra caution to avoid erosion & siltation into storm sewers.
CENTER: Street resurfacing & curb repair can themselves become stormwater problems with poor planning & preparation.
BOTTOM: Minor repairs pose special problems: loose paint chips carry mercury & other poisons to lakes & rivers.

levels by a factor of 3.8; 5) the water quality in the upper basin (Cedar, Isles and Calhoun) has shown a gradual but significant decline since 1958, a period coinciding with the completion of storm sewer construction in the watershed.

If the Diagnostic Study of the lakes weren't enough bad news, the Minnesota Department of Health issued a fish consumption advisory in May of 1993 for Lake Calhoun due to elevated mercury levels in fish taken from the lake. Walleye catch data revealed all samples contained more mercury than the limit established by the Health Department for women of child-bearing age and children below the age of 6. The advisory is similar to one established for Lake Harriet in 1989 due to high levels of mercury and PCBs found in fish tissue.

But the Parks Department, which controls the Chain, though not the stormwater that pollutes it, waffled on the severity of the situation. At first Parks, speaking of the 1991 tests, suggested there was no cause for alarm. Parks insisted water quality in the Chain "had not changed significantly in the past 20 years."

Yet many daily lakes users went public with their concerns, saying they had lost confidence in the Park Board's ability to manage the city's precious amenity. At two well-attended meetings they criticized the policy of "managed decline" and wondered aloud why there could not be a "restoration" to the fabled realm of 1927?

A new Lake Water Quality Management Citizens Advisory Committee was convened in July of 1993, with Committee members toiling to

'93 ADVISORY AMBROSIA

- 1 permanent Citizens Advisory Committee
- maximum sweeping effectiveness
- reduce de-icing chemical
- 300 fewer geese
- citywide construction ordinance
- 5-fold increase zooplankton
- 1 ton aluminum sulfate
- 1 name change

Beat poor construction practices thoroughly. Combine with one stick better sweeping, fewer goose pellets, and reduced salt. Fold in fish management and alum to bind phosphorus to bottom. Add permanent resident input and restored name for increased concern and clarity.

reconcile the popular sense of emergency with the narrow Park Department agenda. Of particular importance to the Committee was the subject of wetland reconstruction to intercept and clean stormwater. Wetlands, even rebuilt ones, are typical stormwater treatment rated highly by NURP.

Wetlands, Park officials said initially, weren't in the picture. There was no room for them and to construct them now would take too many homes. Parks refused to support the small but original Ewing wetland next to Cedar Lake, allowing it to go under the development knife. When the Advisory Committee suggested a wetland restoration on the northwest shore of Lake Harriet to deal with the 20% of its phosphorus pumped from the upper chain, Parks proposed instead spending $1 million to pipe dirtier Calhoun overflow under Lake Harriet to the Mississippi River via Minnehaha Creek.

But Committee members were overwhelmingly in favor of wetlands to trap and filter out (50% effectiveness) offending phosphorus. Both as a scrubber and as a bridge to a more idyllic past, the prospect of wetlands working along the shore united Committee members and became a key element of its recommendations.

Disagreement also surfaced on the tricky question of overall goals for the Chain. Parks declared the fate of Isles beyond the charge of the Committee, and offered limited objectives for Cedar and Calhoun, moving them from eutrophic to "slightly" mesotrophic; Lake Harriet would be "protected" at its current pollution level. A short-term (5yr) target of 10% pollution reduction, and a long-term (10yr) target of 20% pollution reduction for all water quality management activities was offered by Parks.

These objectives were fairly modest judged by 1927 standards

Parks eventually committed to three constructed wetlands. The first, Cedar Meadows, was built on a previous wetland along the southwest shore of Cedar Lake in 1995. Its effectiveness has been hindered by loss of first-year plantings, invasion of carp and excessive rainfall. A second, at the southwest corner of Calhoun, is currently under construction, while a third, at the northwest corner of Harriet, is possible.

and serve to underline the role played by the public in both creating the problem and remedying it. When it released Committee recommendations in April 1994, the Park Board admitted Cedar and Calhoun were "in trouble" and "rapidly becoming worse." But government agencies such as the Park Board and Public Works don't have the resources to overcome the 'stormwater stalemate,' so they can't be relied upon to do the job alone.

Intercepting polluted stormwater is far less efficient than "source control" - removing pollutants from the runoff stream at the outset. Government is reluctant to force the necessary resident compliance with the whole range of home best management practices that could eliminate the problem before it hits the streets. Such action must come from desire — a desire to clean up by the individual polluters themselves.

The promise of a permanent Water Quality Citizens Advisory Committee (WQCAC), supposedly an EPA requirement, soon fell victim to stalemate. The Committee met for two years beginning in 1994, hoping to become involved in a variety of actions (see sidebar), from crafting the proposed erosion control ordinance, to taxing plastic 'throw-aways.' But Public Works insisted there would be no commitment of staff time or money to the Committee until proposals for a new 'stormwater utility tax' and a multi-million dollar grit chamber project were ok'd. WQCAC, however, balked at creating a new layer of government with taxation powers, and hesitated to spend $30 million on grit chambers when NURP evidence demonstrated they were largely ineffective (1% removal) against phosphorus, the key eutrophier in stormwater.

Lake 'Calhoun' was re-named for John C. Calhoun, Sec. of War, by a subordinate who surveyed present-day Minneapolis in 1817. The Advisory Committee voted 17-4 to seek a name change for two reasons: 1) 'Calhoun,' as the architect of the Civil War for his career advocacy of slavery, represented a sad example of might makes right; and 2) restoring the lake's name to its native Mde MA-KA-SKA "Lake of the White Earth" (Jos. Nicollet, 1843), would be a first, positive step toward capturing the public's enthusiasm for the hard work necessary to restore the Chain's water quality.

The 'permanent' Water Quality Citizens Advisory Committee (WQCAC), was officially established by city council action on June 5, 1993, and charged to "assist staff and elected officials in developing: a)specific ordinances; b)education materials; and c)development of solutions to severe problems..." relating to stormwater.

At its first meeting, Public Works promised WQCAC would be an "implementation" committee deciding "where can we go, what we can do," as opposed to its predecessor's "planning" function.

When WQCAC responded positively to a Minneapolis Planning Department request for comment on a proposed change in zoning rules to protect 'natural and sensitive features,' staff argued that such comment amounted to "a taking" of private property, constituted "an end run" on Council perogatives and potentially affected Council election chances!

Subsequently, it became known to the Committee that Public Works' first grit chamber on the shores of Lake of the Isles had been 'cleaned' *by dumping the residue back into the lake.* Public Works then complained to the city council that some members of WQCAC were acting as "advocates for the environment." Public Works petitioned Council to limit WQCAC's role to public education. The city council, upset that WQCAC responded to a request for comment outside of channels, concluded there was too much citizen involvement in stormwater oversight, and folded WQCAC "responsibilities" into a staff-dominated existing 'environmental' committee—Citizen Environmental Advisory Committee (CEAC).

As of 1998, few proposals of the 1994 Advisory Committee recommendations had been enacted; the geese range was reduced and Cedar Lake was treated with alum. Alum is a temporary measure, with water clarity improved anywhere from 5-10 years depending on continued phosphorus inputs from stormwater, in-lake circulation from bottom feeding fish, and the number and effectiveness of other BMPs applied.

Study warns lakes in city on way to a scummy future 11/12/70 Star

By BETTY WILSON
Minneapolis Star Staff Writer

Minneapolis lakes soon will be green, scummy and

The city's lakes are "semi-degraded," according to the report made public today of the water quality of seven city lakes.

Metro/State

Minneapolis chain of lakes threatened

Group seeks program to curb pollution

By Dean Rebuffoni
Staff Writer

Warning that the Minneapolis chain of lakes is imperiled by pollution, a citizens' advisory committee is urging city officials to begin an aggressive program to improve water quali-

to promote large blooms of oxygen-consuming algae and other aquatic plants.

More costly measures include restoring drained wetlands on public property along Lake Calhoun and Cedar Lake. The wetlands would filter

treating Cedar and Calhoun with aluminum sulfate. That could phosphorous now contained in sediment on the bottom of those from being released into the wat

Also recommended is a vacuum sweeping program on city st

WATER QUALITY AND BIOLOGICAL INVESTIGATION
OF THE
CITY OF MINNEAPOLIS LAKES
1969 - 1970

CONDUCTED FOR
THE MINNEHAHA CREEK WATERSHED DISTRICT

IN COOPERATION WITH
THE MINNEAPOLIS PARK AND RECREATION BOARD

OCTOBER 1970

EUGENE A. HICKOK & ASSOCIATES
HYDROLOGISTS - ENGINEERS
WAYZATA, MINNESOTA

Water Quality Management Citizens Advisory Committee

THE MINNEAPOLIS CHAIN OF LAKES
A Study of
URBAN DRAINAGE AND ITS EFFECTS
1971-1973

Joseph Shapiro
Principal Investigator
and
Hans-Olaf Pfannkuch

WATER CLARITY
WILLIAM BOUDREAU

A crucial step toward cleaning up our water is to alter habits in simple ways

The most important realization concerning storm water for a city resident is that their property, from front to back lot lines, from alley to curb, is a tributary of all surface waters in the city and downstream. Simply stated, what you put on your property is eaten by fish, swims with you in city lakes, and becomes part of the aesthetic all the way to New Orleans.

Obviously, we can't go back to a pristine, undeveloped local environment, but if we can recognize that such a condition would be optimum for water quality in our lakes, streams

amendment (i.e tilizers). Suffici and nitrogen numbers on the tilizer bags) are ing, healthy la Nature via atmo (it drops from t budding trees, and South D True, lawn t "green up" y they'll also gree Isles and the Mi and you'll have

The second area is the Minn tions Departm they threatened

S-T 8-7-92

Lakes in peril

Group vows to fight storm-water runoff that could bring end to swimming, fishing

By Kevin Duchschere
Staff Writer

Unless quick action is taken, contaminated storm-water runoff may soon make Minneapolis' cherished chain

City lake standards of 1927 proposed 2/17/75

By MARIETTA SMITH
Minneapolis Star Staff

A plan to restore the ment

Minneapolis Lake

Citizens Advisory Committee On Lake Water Quality

To clear Minneapolis' lakes

Broad public effort is needed to halt the spread of pollution

By Carolyn Light Bell

I won't swim in Minneapolis lakes anymore. If I put my head underwater, I break out in a rash.

our lakes. Geese are vectors for swimmers' itch, bacterial infections and viruses.

Eating and defecating, geese are phosphorous pumps. Each

LOTS OF TALK HAS NOT PRODUCED CLEAN WATER

268
MOBIL
SWEEPER

3
BMPs -
HOME RECIPES FOR CLEAN WATER

RECIPE FOR STORMWATER SUPREME

1 qt. Best Management Practices
reduce, reuse, repair, recycle
natural household cleaners

one 5 'X 5' compost heap
1/3 gal. paint savvy
1 c. construction precaution

As homeowners and renters in a large city, we must understand our role in keeping surface water clean. While we may feel powerless as individuals, we belong to a massive permanent encampment on the banks of nearby lakes and streams. Our strength — and our peril — lies in collective action.

Because stormwater sources are so numerous there is no silver bullet which alone can stop the monster. Mechanical solutions such as grit chambers and periodic brush street sweeping (photo, opposite), represent expensive, but not particularly effective engineering solutions. The chambers require street excavations and regular maintenance; brush street sweeping misses fine particles and must be timed perfectly to clean streets before rain and snow arrive.

Nature, redeployed, with deep ponds and wetlands adjacent the lakes to intercept stormwater naturally, is a worthy but sometimes unworkable option in a dense urban setting.

Building traditional multi-stage treatment facilities to clean dirty runoff, or diverting the 'first-flush' - the early plug of poisons moved as thundershowers "clean" the streets - both present tough land-use questions.

Stormwater would have to be slowed to effect sedimentation, and chemically treated at innumerable points to remove phosphorus and nitrogen.

These limitations on large-scale public improvements underline the crucial role home environments play in maintaining water quality in urban areas. Homes represent the environment's best chance: *pretreatment.* If it isn't used, it can't be spilled, dumped or disposed of improperly.

City residences, from alley to curb,

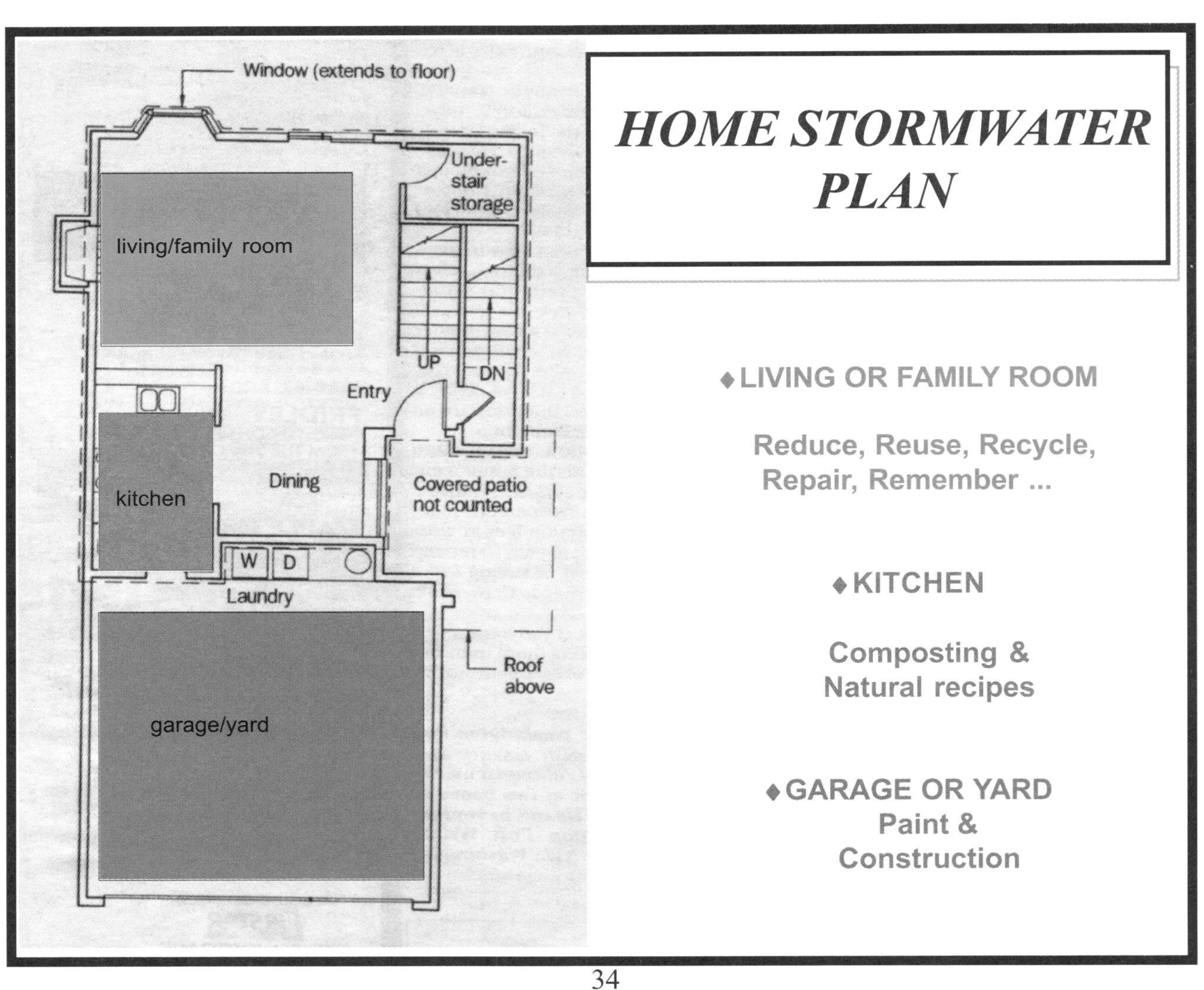

rooftop to laundry tub, are tributaries to all downstream surface waters. Whatever is dosed on the property is eaten by fish, swims alongside and becomes part of the aesthetic from Minneapolis to New Orleans. With understanding and resolve, we can initiate a home stormwater plan to insure that our impact on the environment will be minor.

A successful ***home stormwater plan*** would begin by mapping household areas where long-standing Best Management Practices (BMPs) intercept pollutant loads before they hit home, certainly before they hit the street. Home BMPs are simple recipes for homeowners and renters, along the path to successful source control.

The family room is one of the zones where BMP recipes could achieve significant stormwater clean up. This is the entry point for the 1,000 pounds of trash thrown away each year by every American. About two-thirds of that trash is **packaging**, most of it paper.

During the land-fill crisis of the 1970's, many cities resorted to burning trash (**solid-waste**) as a solution to the problems posed by burying it. The resulting 'Pollution Plume' of contaminated soot and ash from municipal incineration falls over local neighborhoods and watersheds, giving permanent incentive to reducing home wastes.

1) Reduce much of that trash before it comes back to haunt surface water by cutting down on unwanted packaging and mail while increasing use of recyled paper.

2) Return polystyrene packing materials to shippers who can re-use them. Call the Plastic Loosefill Council's "Peanut Hotline" at 1-800-

REDUCE, REUSE, REPAIR, RECYCLE

A rainbarrel
Charities
Commercial outlets

Reuseable containers
Community drop-off
100′ Clothesline

Combine simple procedures like rainbarrel, clothesline and recycling bins for new lake-saving attitude.

Continued on page 39

ORDINARY RECYCLING

(CITY)

MINNEAPOLIS BIWEEKLY RECYCLING PICK-UP

DON'T MIX. PLACE EACH TYPE OF MATERIAL IN SEPARATE BAG. RINSE CANS & BOTTLES, REMOVE LABELS.

aluminum - cans &foil
glass - all colors (remove lids; *must be* separated from other materials)
plastic bottles - detergent, juice, lotion, milk, syrup & pop bottles (no microwaveables, prescription or motor oil)
corrugated cardboard - (flatten & bundle in 3'x3' loads; no pizza boxes)
mail - brochures, cards, envelopes, notes
office paper - office paper
boxboard - dry food boxes such as cakes, crackers, pudding, cereal, potato chips (NO wet-strength beverage boxes! This includes milk.)
tin - seamed cans (clean, remove labels)
household batteries - (place in clear plastic bag; no auto or rechargeables)
magazines - "glossies" (bundle with string or place in paper bag to weigh no more than 20 lbs.)
newsprint - (bundle with string or place in paper bag ... no more than 20 lbs.)
large items - (only 2 per week ... or get voucher*; must be tagged: "For Solid Waste")

- **stoves, refrigerators & microwaves**
- **furniture**
- **carpet** - (must be tied in roll no more than 5' long, 12" diameter, < 20 lbs.)
- **mattresses**
- **washers, dryers & water heaters**
- **tvs** - (a TV screen contains up to 5 lbs. of l*ead*)
- **computers & monitors** - (contain lead, cadmium & mercury)

items 1/2 metal - (cut down to 8')

- **swing sets**
- **clothes poles** - ("T" removed)
- **pipes**

MID-APRIL to MID-NOVEMBER

yard waste - (bag or bundle)
Leaves & grass clippings, bark, twigs, garden plants & straw; bag <40 lbs.
Brush & Prunings - pieces < 3" diameter; bundle < 3' long

Drop Off Transfer Station located @ Mpls Solid Waste & Recycling 673-2917

*Clean-Up Vouchers are required for concrete, construction & remodeling waste, tires, bricks, sheetrock, sand, lumber & asphalt.

HAZARDOUS WASTE RECYLING

(COUNTY)

HENNEPIN COUNTY HAZARDOUS WASTE DROP-OFF*

auto waste & fuel additives, starter fluid, waxes, etc.
floor care products
spot remover
drain cleaner
oven cleaner
commercial furniture stripper/polish
aerosols
appliances
rechargeable batteries - (cadmium is a toxic, bio-accumulative metal)
degreasers/cleaning solvents
pesticides - insecticides, herbicides, rodenticides & fungicides
mothballs
mercury - thermometers, fluorescent bulbs, HID lamps, thermostats, light switches (older "clicks")
nail polish/remover
oil-based paint & paint thinner
wood preservative
driveway sealer
roofing tar
consumer electronics:
- fax modems
- receivers
- stereos
- VCRs
- camcorders
- electric razors
- flash lights
- hand-held vacuums
- portable phones

adhesives
photographic & hobby chemicals
lighter fluid

Info Line: 348-6500

2 Hennepin County Locations:
north - 8100 Jefferson Highway, Brooklyn Park
south - 1400 W. 96th St., Bloomington

*Other County Hazardous Waste Drop-Off Sites:
Ramsey - DYNEX Industries, 4751 Mustang Drive, Mounds View; 773-4488
Washington - Oakdale Hazardous Waste Facilty, 1900 Hadly Ave. No.; 430-6770
Dakota - Gopher Smelting, 3385 S. Highway 749 (Dodd); 891-7011
Anoka - 323-5730 (seasonal)
Carver - 361-1800 (seasonal)

PRIVATE RECYCLING

PRIVATE DROP-OFF

metals - copper , aluminum, steel **- check phone book under "scrap"**

automotive - car batteries , antifreeze, motor oil, oil filters, tires **- see Chapter 7**

styrofoam -

A) **plastic loosefill** - call "Peanut Hotline" at 1-800-828-2214 for a list of recycling merchants near you.

B) **expanded polystyrene** (the molded foam packaging that cushions appliances and electronic equipment) - call the Alliance of Foam Packaging Recyclers @ 1-800-944-8448 or on the Internet @ www.epspackaging.org

poly bags - call the Plastic Bag Information Clearing House toll-free hotline for location of recyclers: 1-800-438-5856 (newspaper sleeves, bread wrappers, self-service produce & grocery bags, dry cleaner bags)

***reuseable* building materials @ ReUse Center**
Hiawatha & Lake St.
724-2608

door knobs	window handles	molding	toilets	lumber
plumbing	light fixtures	cabinets	sinks	doors

CHARITY CONTRIBUTIONS

ARC OF HENNEPIN COUNTY
CALL FOR PICK-UP: 612-866-8820

clothes (clean) - any size, season or gender

household items - pots and pans, dishes, linen, toys, shoes, purses, gloves, hats, nick-nacks, craft items and *small* furniture, i.e., no sofa beds, stoves, etc.

Continued from page 35

828-2214 for a listing of merchants near you. Polystyrene, or "Styrofoam," is a growing presence in stormwater and at the beach. This lightweight (it's 95% air) insulator is not biodegradable. Six billion tons are produced annually with an alarming amount showing up in surface water where its unsightliness and mild toxic impact on invertebrates are a sad commentary on our throw-away lifestyle. Never place them loosely in the garbage.

3) Recycling can be a major factor in cleaner water. Less than a fifth of the nation's products come from recycled materials. Yet virtually every item in the home is recycleable to some government agency, private firm or non-profit charity.

4) Remember that half of the pollution on the planet is caused by the burning of fossil fuels for heat and energy. This burning produces poisonous particles that drop directly on waterways, streets and yards.

Everyone has a role in **reducing home energy consumption**: Switch to compact fluorescent bulbs, they use 1/4 the energy and last 10xs longer. Turn down thermostats overnight, or during the day if no one is home. Turn off unused lights and appliances. Run less water: fully load dish and clothes washers - better yet, put up a **clothesline** (not around a live tree please...forgotten ropes strangle the tree as it grows).

Any solar-powered appliance avoids potential battery spills; rechargeable battery-powered equipment reduces the toxic effects of burning fossil fuels.

5) Reject "disposables" of any kind; they encourage no consequence throwaways that pollute. Absorb spills with durable sponges rather than disposable paper towels.

6) Repair whenever possible. From a runoff point of view, unwanted and discarded items will deteriorate and leak into the environment.

7) Remove your name from junk mail marketing lists, write Mail Preference Service, Direct Marketing Association, Box 9008, Farmingdale, NY 11735-9008; you might also request your address be removed from the lists of the nation's largest mass mailer: Director of List Maintenance, Advo Systems, 231 West Service Rd., Hartford CT 06120-1280.

The **kitchen** too, is the staging area for household stormwater screwups — and a potential area for pollutant savings. Keep in mind that in most cases, kitchen bmps that

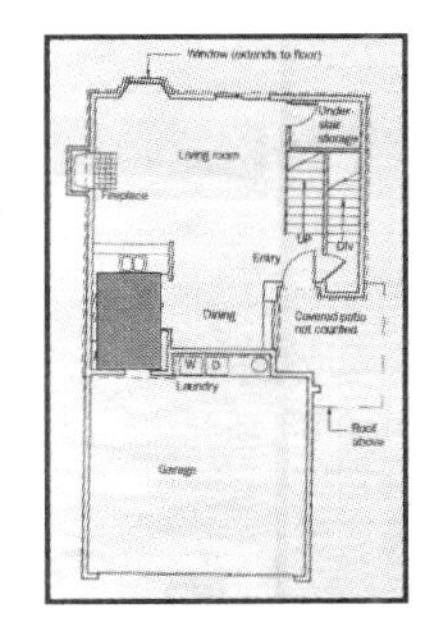

reduce potential stormwater problems, cut spending costs as well.

Begin by reducing packaging. When shopping for groceries, use your own reuseable plastic mesh or canvas bags for the trip home. If you use store packaging, make it paper bags as they are easier to recycle than plastic bags. Where possible, buy in bulk; liquids in concentrate cut costs in half.

When purchasing meat with blood sponges, cut open styrofoam containers, remove sponge, rinse and recycle styrofoam.

8) Reusable containers, rather than plastic bags or wrap, are excellent for leftovers and don't require disposal.

9) Replace synthetic cleaning products with their more benign natural counterparts posing less risk to aquatic environments. Synthetic cleaning products contain regulated industrial chemicals such as concentrated acids or bases, phosphates, chlorine, EDTA, and aromatic hydrocarbons. The Environmental Protection Agency (EPA) estimates that the average household contains 100 lbs. of commercially-blended hazardous substances.

Under no circumstances should hazardous concoctions be poured down household (sanitary) or stormsewer drains!

Down household drains, these com-

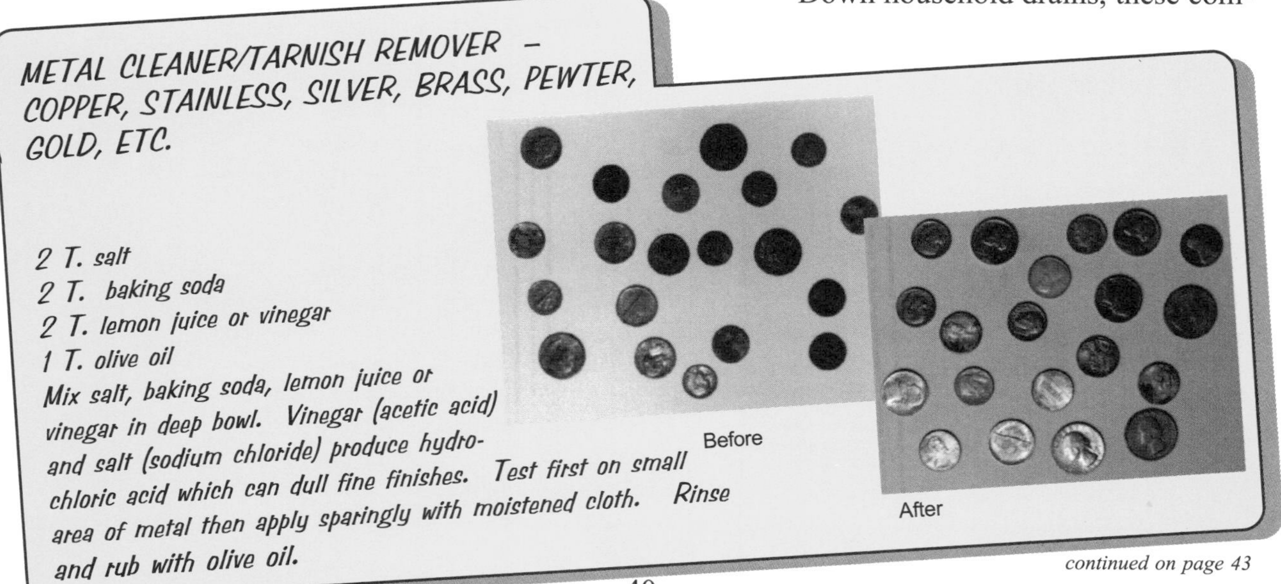

continued on page 43

NATURAL CLEANERS & POLISHERS

CLEAN KITCHEN COMPOTE

1 box baking soda
natural cleaners
1 set plastic containers
1 gal. vinegar
1 recycling container
2 lbs. kitchen wastes

Mix natural miracle cleansers, reusable storage containers, recycling and composted kitchen wastes for simple, pollution-free environment.

GENERAL CLEANER & DISINFECTANT

1/2 c.vinegar
1/2 c. Borax
1/2 gal.warm water

SCOURING POWDER

1/4 c. baking powder
1/4 c. Borax
2 T. salt
Mix, dampen to paste. Scour and rinse.

AIR FRESHNER

1 sliced citrus (lemon/lime/orange) and/or aromatic herbs such as mint, cinnamon, cloves, or heather; 2 c. boiling H_2O. Pour boiling water over citrus/herbs. Let sit over-night (12 hours). Strain. Pour into spray bottle, or add to cleaning formula.

ALUMINUM PANS

Outside, scrub with baking soda. Inside, simmer 1 qt. water with 3 T. cream of tartar for 20 minutes.

COCKROACH COFFIN

2 T. baking soda, 2 T. sugar. Mix and sprinkle near infested areas. A spoonful of sugar will make the 'medicine' go down.

DRAINS

To prevent clogs, pour in 1/2 c. baking soda followed by 1/2 c. vinegar. Let sit 2 hours. Flush with hot water.

CONTINUED ON PAGE 42

NATURAL CLEANERS & POLISHERS

(CONTINUED)

DUSTER/POLISHER

1/4 c. olive/walnut oil, 2 T. food-grade lemon oil, 1 T. vinegar. Mix and apply with soft rag.

FLOORS

1/2 c. vinegar with warm water. Or, Borax and water.

FURNITURE

Combine 1 part lemon juice with 2 parts vegetable oil in small bowl and dab on. (Or, combine 1 part vinegar/whiskey with 2 parts mayonnaise/ olive/almond oil.)

REGULAR OVEN (NOT SELF-CLEANING)

1 c. baking soda, 2 c. water, 1/2 c. washing soda, 4 T. salt. Sprinkle oven with water. Mix baking soda, washing soda and salt; sprinkle inside oven. Let sit overnight. Sprinkle more water and sponge clean. Rinse well.

SUPERBOWL TOILET CLEANER

1 c. Borax. Pour into bowl. Sleep on it (12 hours). Brush and rinse.

WINDOWS '98

Spray with pump bottle mixture equal parts vinegar/water; wipe with newspaper. Dry newspaper, and recycle.

Continued from page 40

pounds are diluted but not totally neutralized. They may corrode pipes - even explode - while passing untreated through the system. Or, burned as treatment sludge, they go briefly airborne before stumbling back onto the streets and down storm drains where they slash at the web of life that maintains lake water clarity.

EPA tests show natural/organic cleaners are as effective as synthetic/petroleum-based products in ability to clean. Organics often cost more but, because they are concentrated, they offer savings on packaging and on a per use basis. They may require however, a bit more of the cleanest burning fuel of all - elbow grease.

Let's face it, many contemporary commercial furniture cleaners, metal polishers and grease-cutters are simply perfumed chemical equivalents of what granny used — with the costly addition of heavily-marketed fancy packaging.

Reducing toxics in the kitchen means reducing the perceived need to introduce them to the home in the first place. Stay away from products labeled *Caution, Warning, Flammable, Volatile, Caustic, Corrosive, Danger or Poison.* Safe, non-toxic products do not require the care and attention in use and disposal that dangerous ones do.

Borax (sodium borate), a mild alkaline compound containing boron, is an all-purpose cleaner, mold-inhibitor and disinfectant.

Baking soda (sodium bicarbonate), the "miracle" chemical, is a mild alkaline abrasive with numerous uses: extinguishes electrical, grease, gasoline and chemical fires (but not wood or plastic), deordorizes (actually absorbs strong acids and bases, the cause of strong odors), cleans (turns fatty acids in grease into soap) and polishes metals, tiles, even teeth.

Salt (sodium chloride), is a mildly abrasive, non-scratching cleaner that kills some bacteria.

Washing soda (hydrated sodium carbonate), sometimes called 'sal soda,' is a milder alternative to chlorine bleach (*paradichlorobenzenes*). Washing soda gets high marks for cutting grease and cleaning dirt but may scratch fiberglass. Chlorinated materials should be avoided as they can create organo-chlorine compounds (such as dioxin) that are stored in living tissue where they can create reproductive havoc.

Vinegar (acetic acid) is a natural substance with many therapeutic uses: combined in equal parts with hand cream, it quickly heals chapped skin; 1-2 tsps. with a glass of water

SLOW COMPOST CAKE

1 bushel grass clippings
2 bushels spent flowers
2 bushels fruit and vegetables, raw or cooked
lots of micro-organisms, insects & earthworms
1/2 bushel sticks and/or coarse material
(OPTIONAL) 1 bushel manure (no dog or cat feces)
4 bushels leaves
5 gal. water
spade or two of soil

Mix ingredients. Top with 1" soil. Simmer@ 150 degrees under pressure, stirring whenever. Screen into wheelbarrow through 3/8" hardware screen tacked to a simple frame. Yield: delicious yard confection.

cures morning sickness, hangovers, and stress (raised blood pH). Added to baking soda, it bubbles dirt away! It also removes gum from fabric and carpet, cuts grease on dishes, seasons skillets and discourages fleas and ticks.

Citrus fruits (including the tomato) are naturally acidic, good for dirt and grease removal, metal polishing and combating odors.

10) Composting would allow us to immediately lop the top from the mound of household pollution. Some 17% of all household waste is **kitchen** debris — fruit cores and peelings, vegetable scraps, egg shells, coffee grounds, tea bags, pasta, rice, cereal leftovers, flowers, shrimp shells, bread, hair clippings - even dust.* Another 17% of household debris comes in the form of **yard waste** — leaves, brush, weeds and garden cuttings. **

By combining green kitchen wastes (nitrogen), with brown yard wastes (carbon) in a compost pile ('compost' means stew), we can produce a rich nutrient cake equivalent to the best balanced organic fertilizer (4-4-4) available. That's because microbes, bacteria, fungi, worms and other tiny creatures regard our plant wastes as food and are willing to recycle them into a valuable soil amendment at no cost.

Since all this material will eventually break down into nutrients anyway, composting allows us to turn it into an onsite resource, building soil and nourishing lawns and gardens with no export to lakes and streams. Backyard composting can lower waste collection costs as well.

Remember too, that food scraps if burned rain down on streets and lakes, while dumped

* Meat, bones, grease and dairy products are not recommended because of odors and unwelcome scavengers.
** Grass clippings may be composted but they are best left on the lawn — no herbicide tainted clippings please!

down a disposal, they contribute enormous amounts of nitrogen to receiving waters since sewage treatment as yet does not require complete nutrient removal. Recent massive shellfish kills in the Gulf of Mexico have been linked to the nutrient-saturated Mississippi River.

Draining nearly half the land area of the United States, Mississippi waters with three times the nitrogen loads of 50 years ago, have punched huge holes, called *hypoxic* or 'dead zones,' into waters off the coasts of Louisiana and Texas. By stimulating an explosion of various algae whose death and decomposition cuts oxygen to the point of suffocation (less than 2 parts per million), excessive nutrients can do unto large areas of the ocean what they've done unto lakes.

While precise control of volume, carbon to nitrogen ratios, moisture, temperature, etc. may speed up composting, you don't have to do any more than throw wetted kitchen scraps and yard refuse into a pile to get underway. It is best however, to make a large pile, 5' wide by 5' deep, to keep internal temperatures high during cold winters.

Turn every summer by forking not-yet-decomposed material into a new pile: harvest the rich soil underneath.

After several months compost will become **humus**, a granulated fertilizer that can be spread 1" deep on lawns, 4" on vegetable gardens, and 6" around trees and shrubs providing soil structure, drainage, and a nutritional spike. Spongy humus worked into sandy soils increases water retention, in heavy clay soils, it improves air circulation.

Compost is also an excellent mulch, suppressing weeds, preventing summer dry-out and winter die-out.

Red worms (*Eisenis foetida*) can be set up to compost indoors year-around. Place a box (kits available) in the basement and just a handful of the smallish worms will happily turn 200 lbs of annual kitchen waste into nutrient-rich casts cleanly and cheaply. When you've done it, start a worm club and call city hall demanding a rebate in your waste collection costs.

Above, large 5'x5' enclosure is preferred for composting. Pictured at right is an example of many commercial compost containers that will do the job.

Every picture tells a story: this one says they're raking leaves into the street ... illegal since 1974! Why not compost instead?

If you chose not to compost, you are legally bound to bag your yard waste for regular sanitation pick-up and recycling. It is **illegal** (Ord. 427270 Leaves, grass on streets, 1960, amended 1974) to place grass, leaves, twigs, and other unwanted yard **debris** into the **gutter.** Please **do not rake leaves into gutter** in anticipation of street sweeping.

Garage and Yard

Paint

A major water quality concern around the home is paint. Paint is basically an *adhesive* coating to protect and enhance the appearance of wood. Paint may be refined from **petroleum** (Oil-base paints, 50% solvents); **water-based** (latex, +/- 5% solvents), or **plant-based** (Livos, APM and Auro), trees, safflower and linseed (0% solvents).

While lead (1978) and mercury (1990) have been removed from interior latex paint, they may still be used in exterior latex paint. It is estimated that half of all US homes still are coated with lead and mercury-laced paint. Lead-based paint has produced an epidemic of learning-disabled children in urban areas; mercury is one of the most dangerous of all heavy metals, poisoning in amounts so small, a dime-sized chip of mercury-tainted paint could produce thousands of birth defects!

Oil-based paint contains several dangerous substances - flammable solvents, ketone, xylene, methylene chloride - added as fungicides and pesticides. Oil-based paint is toxic to hu-

BACKYARD BARBECUE

1 snow shovel/ice chisel
2 plastic tarps
1 downspout
1 gross paving bricks
1/2 doz. upside-down paint cans
large broom

Trap exposed soil, limit salt and sand, and turn downspout toward lawn. Install permanent pervious surface on paths, driveways. Sweep away debris to compost; turn over paint cans to eliminate air leaks and reduce polluted runoff.

mans as well as creatures at the other end of the pipe, whether dried and pealing or poured down the stormsewer. If you have older paint or paint containing toxic compounds, store it as hazardous waste, recycling it to the nearest toxic collection location.

Latex paint, unlike oil-based paint, requires only warm water for clean-up. (Oil-based paint requires petroleum solvents.) Still, paint or water-proof coatings — properly applied — are likely to do less environmental damage than buying wood treated with preservatives.

Paint thinner, a solvent distilled from petroleum and used to clean oil-based paint from brushes, can be reused by screening it through linen cloth or just by letting it set for a few months. The paint will separate and sink leaving the clear thinner on top.

Painter's canvas dropcloths are excellent for collecting dust and spills and will reduce reliance on dangerous chemicals.

To store paint for long periods, pound lid in place (a rubber mallet works best) and turn can upside down. The paint will seal itself and remain mildew free for many years.

Construction

Any major **construction** on your property should be accompanied by measures to prevent soil erosion and insure that the amount and rate of runoff after construction is equal to (or less than) runoff conditions before excavation. The simplest method is to preserve existing trees and grass while reducing slope to slow the rate of runoff.

It is important to keep soils under their vegetative cover as long as possible. Tree roots and trunks can be protected by perimeter fencing reaching to the dripline.

For projects of a few months, hay bales and sediment fabric (aka silt fences) will slow stormwater from disturbed areas but they are

The best drive or walkway construction uses materials which allow for absorption of runoff.

The City of Minneapolis requires that any construction project which disturbs more than 5 cubic yards or more than 500 square feet of soil needs an erosion control permit and an approved erosion control plan before work can begin. (Minneapolis Public Works Department — 673-2406)

susceptible to flooding at above-average rainfalls. They should be installed on downslopes, set within a trench 4" below ground and staked before construction begins.

Other recommendations include:

- Temporarily exposed soil can be covered with plant seedings, mulch, or well-anchored burlap or plastic sheeting.
- Locate soil piles as far away as possible from slopes and curb cuts and cover.
- Temporary driveways should be covered with 3" of gravel to prevent mud from leaving the site and being deposited on the street.
- When building concrete driveways, wash out fines to the side, not down the driveway. On sloped drives, place straw bales at the bottom or divert flow to vegetative cover. While working with concrete or mortar, clean wheelbarrow and other tools over grass or soil where hardened residue may be collected and recycled.
- Never wash finely-graded soil material from your property into stormsewer drains.
- Reduce runoff velocities by vegetating bare soils as soon as possible; divert runoff from exposed soils to vegetation with check dams (straw bales).
- Inspect sediment controls frequently to be certain they are doing their job.
- If using hazardous liquids, prepare a spill response plan and keep a supply of absorbent material such as sawdust or vermiculite nearby for clean-up.
- Cover lumber stockpiles, building materials and metal products.
- Clean up litter at the work site daily.
- See that all site plans requiring building

Wheelbarrows should not be laid down in the street and rinsed!

permits are reviewed by a certified erosion specialist.

Large-scale spills of solvents, fill, oil or other poisons into the watershed can result in legal action and fines.

In winter, avoid using salt or liquid de-icers. They operate within a narrow temperature range, will weaken concrete, and can be traced to increased salinity in local lakes. Use a **chipper** immediately after ice forms or cover lightly with sand. When ice melts, sand must be swept up and stored before it can be washed from property.

Use a rainbarrel to collect and distribute rainwater, a mild acid, helping plants take up vital minerals.

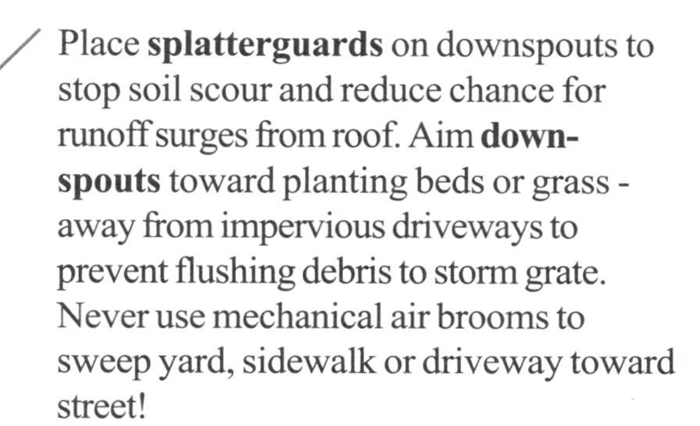

Place **splatterguards** on downspouts to stop soil scour and reduce chance for runoff surges from roof. Aim **downspouts** toward planting beds or grass - away from impervious driveways to prevent flushing debris to storm grate. Never use mechanical air brooms to sweep yard, sidewalk or driveway toward street!

Individuals at some point must take control of their gutters, promptly **sweeping** up garbage, leaves and debris from their territory *before* it has a chance to go down the street grate. Watershed studies suggest phosphorus levels can be reduced 30 to 40% when streets are kept free of leaves and lawn clippings.

PART OF THE 1,000 POISONS WITHIN REACH AT HOME

air freshners - *alkyl phenoxy polyethoxy ethanol*
all-purpose cleaner - *ammonium hydorxide, alkylbenzene sulfanate*
bleach - *chlorine; dioxin, ozone-depleting gases*
drain opener - *hydrochloric acid, potassium hydroxide, sodium hydroxide*
dishwater detergent - *chlorine bleach, phosphates*
floor finish - *ammonia, diethylene glycol, petroleum distillates*
metal cleaners - *ammonia*
oven cleaner - *sodium or potassium hydroxide*
paint, thinner - *toluene, chlorinated hydrocarbons*
polish - *petroleum distillates, 1,1,1-Trichloroethane, phosphoric acid*
T-bowl cleaner - *chlorinated phenols, hydrochloric acid, oxalic acid*

4
THE END OF LAWNS AS WE MOW THEM - LANDSCAPING TO ELIMINATE RUNOFF

RECIPE FOR NATURALIZED LOT

1 meandering bark path
wildlife nooks
perimeter "habitats"
benches for contemplation

shaded areas
ponds/catchment basin
special plantings
steep slopes (1′ vertical to 2′ horizontal)

Carve walking paths between specialized yard areas using porous materials instead of concrete. Use ground covers and mulches for shaded areas, bare soils and property edges. Level slopes with stone bulwarks to slow runoff; trench sidewalks and alleys to catch runoff for slow release. Woodland, Meadow or Water Gardens provide interest and reduce chemical dependency.

Today's landscape ideal combines a narrow selection of plants with a rigid and formal scheme for their placement about the property. We've made the house the focus of our yard, surrounding it first with a deep, flat border of lawn, and secondly, an inner ring of common shrubs and 'aliens' — non-native flowering plants.

According to this view, weeds (a flower at the wrong address) spread hayfever and ruin the effect, while concrete requires little watering or mowing and remains in bloom year around. Vast stretches of imported, sterile grass conveys success, even nobility while local wildlife is perceived as a threat. Poisons and conformity are good; diversity and critters are bad.

The current percentage of land use turned over to these conveniences of modern living comes at enormous cost. The 'tax' we pay for a steady infusion of expensive and harmful chemicals is chaos among the native plant communities and a more turbulent, powerful runoff. It's the little things producing minor effects at first, which later on fragment habitat and interrupt food chains, diminishing the quality of life for all.

Americans spend $25 billion annually on lawn-care products, with the average suburban lawn using 10 times the pesticides of a similar sized area of farmland. While a gasoline lawn mower emits the same level of pollution in an hour of operation as a car driving 350 miles.

Monoculture plantings hurt soils, depleting them of minerals, and inviting disease. Research shows that diverse plant communities - **woodlands, meadows** or **prairies** for example - are better able to endure the stresses of extreme weather and insects. As species fill roles that complement each other, flowering and dying at different times, soil is less likely to dry out. There is seldom any erosion or flood of pollution into the stormwater river from such sites because water runs slowly and pollutants are absorbed in decaying vegetation. Also, these natural **Xeriscapes** reduce water needs 50-80% over

Background yard consumes 10,000 gallons of water per year. Yard in foreground requires no irrigation.

intensive turf cultivation.

Urban homesteaders can make landscape choices to achieve rewards that can't be found in grass — without sacrificing water quality. Imagine a central space not the house -- be it a patio, deck, statuary, pond, or flower bed -- and border it with a natural diversity. By restoring portions of their property to 'eco-gardens' — low-water, low-maintenance, self-fertilizing, pesticide-free areas - residents can map their property to create water absorbency for capture and on-site treatment of potential runoff.

In addition, your **naturalized** lot could become a link in a wildlife corridor system serving bird migration, native plant propagation and beneficial insect development. (Contact the National Wildlife Foundation's Backyard Wild-life Habitat Program @ 1-800-822-9919 for more information.)

Listed here are several Theme Gardens from many possibilities that spring up around a commitment to reducing lawn dependency and one's level of polluted runoff. All plant listings are samples from larger collections that can be explored by consulting landscape architects, libraries, university publications and nursery growers. We've resisted the call to place a 'wildflower' garden among the recipes because we've tried to sow native flowers among the lists while limiting 'alien' choices.

Begin with slopes. Any slope that rises at a rate higher than 1' vertical to 2' horizontal, is steep for turf grasses, difficult to mow, prone to erosion, weeds and runoff. However, reinforced with rock, trees, native plants, grasses and groundcovers, this situation offers a springboard to multiple delights.

Use soft, curving lines and perennial

RECIPE FOR YARD DIVERSITY

3 evergreens
3 deciduous trees
3 fruit trees
2 dozen timbers
100 perennials
5 50-100# boulders
6 shrubs
100 stepping stones
1 lb. prairie flowers
10 gal. small round rock

To create beautiful wildlife pockets, ornithological oasis, and improve pest control, mix ingredients into areas of contrasting height, mass and color. Let stand for 25 years.

'drifts' to connect different uses of the property, planting smaller flowers so they won't be concealed by larger ones. Group similar colors in clumps of 10-15 bordered by green shrubs.

Choose local genotypes of selected plants, i.e., from stock within a 100-mile radius (available in some nurseries). Do not raid local woodlands!

Woodland gardens utilize the single most dramatic feature of residential property and the essential key to clean water: trees. Tree roots hold the soil together, their leaves return nutrients to the ground (at a rate twice that needed to maintain the tree itself) and shield bare earth from the impact of falling rain.

Trees are home to numerous creatures, providing food, shelter, protection and a place to raise the next generation. They insulate homes, cutting northwest winter winds and shading during summer for a 10% annual energy savings in heating and air-conditioning. In addition, a large tree can reduce noise and filter 50-60% more air pollution than a smaller tree can. Trees provide fruit, visual appeal, and a mature tree is said to

RECIPE FOR WOODLAND GARDENS

Trees:
Gray birch (Betula populifolia) 40′
Maple (Acer spp.) 40′
Hawthornes (Crataegus spp.) 20-30′
Oak (Quercus spp.) 50′
Crabapple (Malus spp.) 25′
Beeches (Fagus spp.) 40′

Understory:
Winterberry (Ilex verticillata) 10′
Gooseberry/currants 8′
Bracken fern (Pteridium aquilinum)
Red-osier dogwood (Cornus sericea) 15′
Blueberries (Vaccinium spp.) 4′

Flowers:
Goldenstar (Chrysogonum virginianum)
Wild ginger (Asarum canadense)
May apple (Podophyllum peltatum)
Early Meadowrue (Thallictrum diocum)
Trout Lily (Erythronium americanum)
Solomon's seal (Polygonatum bifforum)
Columbine (aquilegia canadensis)
White Trillium (Trillium grandiflorum)
Rue anemone (Anemonella thalictroides)

Stir up notions of mass conformity. Mix native species of differing heights and textures. Yield not to the impulse to tidy up. Add shade, peat moss, leaf mulch, brush piles, deadfalls & stumps and benign neglect for natural water cleansing.

add thousands of dollars to real estate value as well.

Trees don't have to mean leaves, i.e., muscle-ache. Evergreens keep their leaves, enlivening the winter scene.

A true woodland garden is allowed to go wild, sort of. In a native woodland there aren't any weeds to weed. Its composition, too, must be natural, trees and flowers growing, and leaves falling, as they would in a true woods. But aggressive invaders such as buckthorn, purple loosestrife and others should not be allowed to take root. Nor should a random placement of common landscape trees and turf be seen as a 'woodland.'

Large-flowered or White Trillium
(*Trillium grandiflorum*)

Woodland wildflowers tend to be subtle bloomers with delicate petals and short-lived flower heads. They begin to appear in early spring, pale and ghostly in the chilly sun. As summer builds a leafy canopy over-head, the quiet twilight of the climax forest is interrupted by the startling beauty of a late bloomer twisting in a shaft of soft light.

Large-flowered Bellwort
(*Uvularia grandiflora*)

Be sure to include several varities of native fern in your garden to add richness and texture to your landscape.

Meadows are a combination of grasses and wildflowers. They form a transition zone between relatively arid prairie communities and wetter woodlands. Small city plots may not attract red-tailed hawks and white-tailed deer, but chances are you will see hawk moths, deer mice, and swallowtails.

Prairie jam offers stands of hardy, deep-rooted natives suited to dry conditions. Though their flowers are not as large as hybrid nursery varieties, their bright petals offer a spectrum of wild colors all summer long. Once established, they require virtually no maintenance. Avoid

RECIPE FOR MEADOW GARDENS

Evening primrose (*Oenothera biennis*)
Obedient plant (*Physostegia virginiana*)
Purple coneflower (*Echinacea purpurea*)
Sweet grass (*Hierochloe odorata*)
Prairie dropseed (*Sporobolus heterolepis*)
Cardinal flower (*Lobelia cardinalis*)

New England aster (*Aster novae-angliae*)
Pearly everlasting (*Anaphalis margaritacea*)
Coreopsis (*Coreopsis palmata*)
Prairie cordgrass (*Spartina pectinata*)
Bergamot (*Monarda didyma*)
Blue false indigo (*Baptisia australis*)

Sow seeds @ rate of 7-10 lbs. per acre in spring or early fall on well-prepared bed free of competitors. Weed-free plot requires smothering and baking existing seeds and plants with clear plastic or window glass covering for 3-4 weeks. Water plot to germinate surviving seeds; remove seedlings and till soil about 1". Or, remove soil to depth of 4" and add sterile soil to plot. Lightly rake 1/8" deep into soil. Roll plot to ensure contact, water regularly but sparingly (15 min.) for 6 wks. Cut back garden height to 6-10" first season; mow to 3" second spring.

"instant" mixes which are generally over-priced and don't reseed well.

Prairie flowers take three years to become fully established. Grasses and weeds will grow first; pick the aliens, mow the grass once a summer and provide adequate moisture.

The fall prairie has lost its bloom but retains the thick texture which prevents erosion into nearby waters.

Butterfly preserves provide a backyard visual feast for the entire family. The butterfly is more than a winged firecracker, a brightly-lit cheerleader for Team Insect. It's also a research vessel, a biological probe into how far we can push the environment before it pushes back. In recent years, many of these flaky little treasures have disappeared as the small plots of land and unique plants they depend upon have been plowed under for yet another shoe store or theme restaurant.

But we can fight back. We can restore this rhapsody of color and preserve a portion of our own capacity for awe in the bargain by simply ripping up a chemically-drenched section of monotonous green lawn to make a butterfly garden. And by picking the plants we can actu-

RECIPE FOR PRAIRIE JAM

Big bluestem (Andropogon geradii)
Switch grass (Panicum virgatum)
Queen-of-the- prairie (Filipendula rubra)
Indian grass (Sorghastrum nutans)
Leadplant (Amorpha canescens)
Tall blazing star (Liatris pycnostachya)
Goldenrod (Solidago ssp.)
Side oats grama (Bouteloua curtipendula)
Round-lobed hepatica (Hepatica americana)

Gray-headed coneflower (Ratibida pinnata)
Common evening primrose (Oenothera biennis)
Little bluestem (Schizachyrium csoparium)
Prairie Smoke (Geum Triglorum)
Black-eyed Susan (Rudbeckia hirta)
Prairie Coneflower (Ratibida pinnata)
Pasqueflower (Anemone patens)

Establish a prairie garden with preparation similar to that of the meadow garden (above). Prairie gardens require 2-3 years to become established; weed thoroughly to eliminate aggressive competitors. Add full sun .

ally select which of our tiny neighbors we want to fascinate us! Once established, a butterfly garden — which is little more than a reflection of native species — requires, indeed begs, for low maintenance.

Butterfly gardens of course, must be built to accommodate not only the fragile perfection of the adult colorwheel, but also the delicate though prodigious appetites of the teenager of the species, the caterpillar. These mobile stomachs consume enormous amounts of green leafy vegetables — 100,000 times their birthweight — on their short path to adulthood.

Silver-spotted skipper on white yarrow

What this means is that to establish a butterfly colony, one needs to select different plant types, some that offer nectar to adults for flight fuel, and others which provide protein for fast-growing bodies. The best combination maximizes the limited splash of solar energy which is a northern summer by offering several varieties of summer-long blossoming annuals and hardy perennials.

Butterflies, being bugs, are cold-blooded and require internal temperatures of 80-90 degrees to become active. Not suprisingly, butterflies and butterfly plants tend to be of the sun-loving variety with senses and behavior tuned over millenia. Butterflies for instance, perceive preferred colors of *purple*, *pink*, *yellow* and *white* in the ultraviolet spectrum as an explosive pulse of light — a sensitivity not shared by humans. What appears as aimless flight at first glance, is in fact a dangerous, fast-paced, search-and-enjoy mission to a dramatic realm of the planet we will never see. And the purpose of these exotic airmen is nothing less than the salvation of the earth itself!

Nearsighted, they must 'scratch' a plant with tastebuds on their feet, in order to detect the specific tastes

RECIPE FOR NORTHERN BUTTERFLY GARDEN

Anise hyssop (Agastachefoeniculum) 2-3'
Astilbe /garden spirea (Astilbe x arendsii)
Black-eyed Susan (Rudbeckia)
Blazing star/gayfeather (Liatris)
Butterfly bush (Buddleia) 1-2'
Butterfly Weed (Asclepias tuberosa) 2-3'
Carrots (Daucus carota sativa)
Common Milkweed (Asclepiassyriaca) 4'
Coreopsis/tickseed (Coreopsis auriculata)
Cosmos (Cosmos bipinnatus)
Crown vetch (Vicia spp.)
Daylily (Hemerocallis)
Dill (Anethum graveolens)
Dwarf bush honeysuckle (Diervilla lonicera) 3'
Gas plant (Dictamnus albus)
Joe-Pye Weed (Eupatorium fistulosum) 4'
Lantanas (Lantana camara)
Marigold (American or African)
Mexican sunflower (Tithonia rotundifolia) 4-6'
Mints (lemon, spear, etc.)
Nasturtium (Tropaeolum majus)
New England aster (Aster fikartii)
Parsley (Petroselinum crispum)
Pearly Everlasting (Anaphalis margaritacea)
Pincushion (Scabiosa caucasica atropurpurea) 3-4'
Purple coneflower (Echinacea purpurea)
Queen Anne's lace (Daucus carota)
Running serviceberry (Amelanchier stolonifer) 2-5'
Thistles (Cirsium) 3'
Verbina (Verbina rigida)
Wallflower (Cheiranthus allionii)
Wild bergamot/bee balm (Monarda fistulosa, didyma)
Wild sweet William (Dianthus barbatus)
Zinnia (Zinnis elegans)

Combine with 3 months sun to produce mourning cloak, painted ladies (pictured above), tiger swallowtail, monarch, comma, red admiral, great spangled fritillary and numerous other flying insects with marvelous sequined wings. Yields hours of fascination.

needed for their survival. While they may devour gargantuan amounts, butterflies have evolved to use only certain plants whose sour, mildly toxic juices, once ingested by the larval form, protect the adult from predation as well. It's a rare bird that having attacked a butterfly raised on dill or elm or milkweed, will repeat the distasteful experience.

In return for this life-giving service, butterflies become tireless pollinators on the plant's behalf, spreading reproductive spores so that the beauty and ingenuity of one generation is reborn in its children.

The butterfly gardener must respect this evolutionary track and stick to native varieties of best-loved flowers and shrubs instead of the gaudy "doubles" and other hybrids which lack attractive odors, potent nectars or protective flavors.

A caterpillar house requires a simple frame, wrap-around screening and top to protect larva from ferocious wasps and hungry birds.

A similar concern for the life stages of butterflies dictates other features of the butterfly garden: 1) don't clean up — some butterflies overwinter in leaf litter; 2) provide shelter from the wind for these occasionally exhausted airboarders; 3) absolutely no pesticides!! There may be 'bad' butterflies from one point of view but they can be handled by hand, without resort to indiscriminate killers; 4) drink in the form of a can or bowl filled with sand and saturated with water; 5) a large, flat rock in your garden makes an excellent basking platform; 6) smear over-ripe fruit (banana, peach, etc.) to fence post as a snack, or soak cotton ball in sugar-water (10 parts water 1 part sugar) and affix to stem or twig with paperclip; (7) you may ensure the arrival of butterflies by obtaining eggs from a mail-order biological supply house in your area; and (8) butterflies are all about vanishing habitat - if you don't create it, they will not come.

If butterfly gardening weren't reward enough for a clean water recipe, the flip side -- moths-- come free. That's right.

Without the swiftness or warning colors of butterflies, moths have become children of the night. Thrill to the mysterious flutterings of this spectrally silent species. Stare agog at the bizarre architecture of the giant 6-inch silkmoths. Marvel at their plush, muted, fascinating bugness. Quake at the diversity of mothlicism — over 10 times as many species as butterflies. Stake a sheet to the garage, light it with a bulb or 9-volt flashlight, and guess whose coming to dinner?

While most adult moths aren't equipped to feed, some are. As caterpillars they depend on native flowers and trees — often the same chosen by the state's 163 varieties of butterflies. Set up a feeding platform and equip it for close-up photography and make a change in your life as different as night and day. Because there are so few professionals in the field, backyard observations of these nocturnal nomads can make contributions to science equal to those of amateur astronomers.

Bressingham Blue Hosta

Water gardens soothe with the magic sounds of moving water. The quintessential element, water is bound to attract wildlife. And water gardens are certain to attract wildlife watchers as well.

Plan to make your pool deep enough — 3-4' — to accommodate fish providing them a cool, safe haven from summer heat and marauding raccoons.

When the hole is finished, line it with soft sand and prepare a thick plastic liner to cover the contours you've created. Either give away the excavated dirt or use it to construct raised plant beds.

Periwinkle

RECIPES FOR FLOWERING GROUNDCOVER SPREAD

Golden carpet (Sedum acre) 6″
Common periwinkle (Vinca minor) 10″
Crown vetch (Coronilla varia) 24″
Hosta (Bressingham Blue) 2-3′
Hosta (Golden Tiara) 10″
Creeping phlox (Phlox subulata) 6″
Fragrant rock cress (Arabis Caucasica) 10″
Speedwell (Veronica longifolia) 4″
Barren strawberry (Waldsteinia fragarioides) 4″
Wild/European (Asarum canadense) ginger 3-4″
Bishop's weed/goutweed (Aegopodium podagraria) 12″
Perennial geraniums (G. macrorrhizum) 12″
Japanese spurge (Pachysandra terminalis) 6″
Lily-of-the-valley (Convallaria majalis) 6″
Tawny daylily (Hemerocallis fulva) 3′
Albo-marginata (H. undulata) 2′
Soapwort (Saponaria) 6″
Ajuga/carpet bugleweed (Ajuga reptans) 8″
Violets (Viola) 6″
Pachysandra

Mix to solve problems of shady areas, slopes, poor soils and erosion. Will spread so give them plenty of room.

Pachysandra

RECIPE FOR FRAGRANT FLOWER AROMA THERAPY

Heliotrope (Heliotropium) 'marine' or 'vanilla'
Four o'clocks (Mirabilis jalapa) 3'
Sweet-scented nicotiana (N. alata) 3-4'
Mignonette (Reseda odorata) 12"
Ornamental sweet peas (Lathyrus odoratus) 2-8'
Sweet alyssum (Aurinia saxatilis) 'basket of gold' 9"
Columbine (Aquilegia) 'McKana's giant' 2-3'
Evening Primrose (Oenothera)
Evening Stock (Mathiola incana)
Night Jasmine (Cestrum)
Rain Lily (Cooperia)

RECIPE MOON GARDEN GLOW

Moonflower (Ipomoea alba)
Hibiscus
Bleeding heart (Dicentra spectabilis) 'Alba'
Feverfew (Chrysanthemum parthenium)
White Sonata cosmos
Pearly Gates morning glories
White lupine
Clematis
Delphinium
Impatients
Misty White nigella

Mix all-white varieties of each for haunting moonlight experience. Not for the romantically challenged.

Build a shallow shelf at approx. 1' for potted aquatic plants. Aquatic plants need 6 hr. of sun per day. The pots will remain submerged throughout the summer but can be taken indoors for winter. (They can be trained to overwinter outside by constructing a two-tier 'cold frame' of hardware cloth and heavy gauge poly sheeting wrapped around two-by-fours).

It's a good idea to add a waterfall at this time not only to oxygenate the water for your finny friends, but also for its magic rhythms. This will increase the cost as water pumps can be a bit pricey, but there isn't a great deal more work involved. Place the pump in a concealed corner of the pond. Run flexible PVC pipe from the pump in a 8" trench to a flat rock on the other side of the pool. Allow the water to spill off the rock about 1' above the intended water level. Disguise the pipe outfall with rock and vegetation. Your submersible pump comes with a waterproof cord which you can connect to any outlet. Viola, water music!

BASIC RECIPE FOR WATER GARDEN

20′ PVC Pipe
5 bags concrete
50 shovelfuls of dirt
30# ledge rock
2 Arrowhead (Sagittaria graminea)
Water hyacinth (Eichhornia crassipes)
Water lilies (Nymphaea):
(N.James Brydon) red, fragrant
(N.odorata gigantea) white, fragrant
(N. Charlene Strawn) yellow, fragrant

2 1/2 HP pump
20′ heavy-duty pool liner
4 goldfish
1 gal. pebble rock
Blue flag iris (Iris versicolor)
Cattail (typha latifolia)
Lotus (Nelemabo lutea)
1/2 doz. Algae-eating snails
1″of fish per 5 gal. of H_2O

Dig out basin to form mini-watershed. Add liner (or concrete), pipe and pump. Fill with water and let stand three days to disipate chlorine. Place aquatic plants in thick pots and top with pebble rock. Create rocky waterfalls, build bridges and stock with fish to keep free of mosquito larvae and algae.

Fill with a few inches of water before straightening the liner. When you've completed filling the pond, let stand for several days to detoxify the water before adding fish. Since evaporation is a constant problem, you may wish to purchase a filter or other water purifiers.

Numerous plants are available but when using cattails, place them in a separate container buried next to your pond rather than in it as cattails spread rapidly. Fish generally have to be overwintered in basement tanks.

While some water gardeners are quite fastidious about draining and cleaning their pond, this isn't necessary if you like a more natural look. If you allow muck to grow on the bottom it would be wise to invest in a water filter and submergent plants to cut down on the growth of algae.

Japanese gardens are noted for their deliberate manipulation of basic elements - stone, water, plants and statuary - to create focal points, provide quiet contemplation, or even tell a story. For instance, stones of varying sizes might stand for the relationship of man to the earth and larger universe. Or the Japanese island homeland could be represented with a pagoda surrounded by a "sea" of small round rock raked into wave-like patterns.

Water introduces sound and movement symbolizing the rhythms of life. Water spells purity and mythically represents a barrier to evil. Islands and bridges are popular for this reason.

Plants are selected not for riotous color

as in a Western garden, but rather for texture, fragrance and balance. Trees, especially pines, are favored as symbols of contact between humans and the wild as they murmer a promise of longevity through their thin pungent needles.

Statuary provides interest and the human element. Represented by stone lanterns, fences, gates, arbors, swings and tea houses, constructed objects are optional though intriguing additions offering a sense of time and mystery.

Japanese-style gardens offer runoff protection due to their lack of chemicals. They also provide serenity, individuality and a pleasant break from the monotony of trimmed turf.

Boulevard gardens are a special case. Boulevards are crucial actors, perimeter soldiers on the edge of the stormwater torrent. Intensive cultivation of boulevards either for flowers or vegetables must be handled with extreme caution. Boulevards comprise the last line of defense before pollution pours off sidewalks, slopes, and lawns plunging into concrete spillways aimed at our lakes and streams. Too often they are bare, neglected, or treated like trash containers.

Planted mounds or berms between sidewalk and street, may slow runoff from sidewalks, but they speed runoff from the boulevard, often with disasterous effects. Here turf makes more sense than mounded soil. If anything, such

RECIPE FOR JAPANESE GARDENS

2-3 large boulders
several yards plastic sheeting (punctured)
hollowed stone or Tsukubai
ornamental tree or shrub
wide-tooth wooden rake
several yards small crushed granite or gravel
moss
1/2 ton of sand

Place large rock, hollowed stone, moss and shrub in central location. Lay plastic fabric and cover with sand. Top with smaller rock or gravel. Swirl with wide-tooth rake. Sit and marvel.

RECIPE FOR BOULEVARD 'GARDENS'

edging or trench device
large part daily vigilance
40' hardy turf
1 packet of seed, covered & staked
slight excavation
cautious mowing
minimum soil disturbance
restricted flower/vegetable beds

Excavate boulevard below sidewalk and curb. Seed with turf grass; reseed as bare spots develop. Aim mower gate away from street. Restrict flower plantings within 2"x6" borders.

perimeter areas should provide "negative drainage," that is, capture runoff. Catchment areas to hold runoff can be constructed by trenching both the inside and outside edges of boulevards next to driveways and sidewalks.

The back of the lot adjacent the alley should be treated as if it too, were a boulevard, one adjacent a stormwater canal. While individual actions may seem insignificant, policing some 1,000,000 daily trips across the metro's boulevards is hardly a trivial pursuit.

Two examples of bad boulevard practices:

Unrestricted gardens, left;

Uncared-for boulevard, right.

Direct downspouts at the lawn rather than at impervious surfaces (left).

Bad trends: house to alley is entirely impervious (right).

Gutter scene: just two weeks after the biannual municipal sweep!

These flo-bulbs are heading for a fall and breakage. Don't put poisons in the path of stormwater ... you may as well be fishing with dynamite.

Low-maintenance lawn (foreground) in summer dormancy.
High-maintenance green lawn in background.

> **That thing is called 'free' which exists from the necessity of its own nature alone, and is determined to action by itself alone. That thing is called 'compelled' which by another is determined to existence and action in a fixed and pre-scribed manner.**
>
> **- Spinoza**

5

I FOUGHT THE LAWN - AND THE LAWN WON

NEW LAWN POP-OVER

soil test
1 rototiller
1 garden rake
10 million (select) grass seeds
20 burlap bags
50# balanced organic fertilizer
pH neutralizing act
4 weeks
1 good drenching

Use soil test to determine missing organic components. Supply organics at rate of approximately 1# per 1,000 sq. ft. (30′ x 30′ section). Till into soil. Rake out weed stems, roots, seeds and thatch. Before reseeding, water. Wait two weeks. Allow emerging weeds to grow to 6″. Pull weeds, retill or both. Reseed with turf grasses. Cover with burlap to prevent seed from drying out. Water lightly 2-3 times per day. Yield: vigorous turf growth, invisible weeds.

Every year the average American household devotes two week's worth of wages to the care and luminosity of that 25 million-acre ribbon of civic merit, the turf grass lawn. Anthropologists suggest the love of turf grass stems from a communal ideal: a flat, safe, sunny paradise after the dim ancestral terror of the trees. Others analyze the phenomenon of the seamless lawn carpet as recent social ritual whereby the lawnmower and fertilizer bag stand in for missing self-esteem. Whatever its source, the case for turfgrass is persuasive:

REPORT CARD FOR LAWNS

A	Traps rainwater, limiting runoff to 10% of rainfall;
A-	Absorbs greenhouse gases that contribute to global warming;
B+	Provides uniform, resilient platform for outdoor activities;
B	Reduces glare and noise, cools local environment;
B-	Produces daily oxygen needs of 100 people;
C+	Feels good on toes;
C	Absorbs dust;
C-	Serves to set off formal landscaping;
D	Creates choice of low maintenance or high-maintenance varieties.

HIGH MAINTENANCE LAWNS

Unhappily, demand for rapid growth, uniform texture, and evergreen curb appeal, can lead to a **high-maintenance** lawn program which overwhelms sound ecological practice, endangering long-range soil productvity and downstream water quality. High-maintenance lawns pose dire consequences for all of God's children with over half of lawn chemicals <u>misapplied</u>.

The domain of the golfer, chemical giants and the retentive suburbanite, the high-maintenance lawn is the Steven King of ground covers. Fast-growing grasses are fragile as infants and must be watered, cut and fed in precise, regular increments. Intensive grass cultivation requires 3-4 times the labor, twice the water, much higher costs and poses a significantly greater threat to surface water.

Deluded homeowners and their commercial lawncare cronies now spread more pesticides and fertilizers per acre than U.S. farmers. Evidence suggests that between church and golf during the lazy, hazy, crazy weekends of summer, the urbanite goes for the green, seeking instant lawn parity with a good hosing of unnecessary

S-S-S-SMOKIN'!!

HOT CHEMICAL GOULASH

automatic sprinkling
zillions of dead soil microbes
tad rotenone
clippings removed

weekly mowings
pinch of 2,4-D
$300-$500
1" - 2" leaf height

Liquid Nitrogen fixes: Urea 45% N or Ammonium Nitrate 33% N

Mix 25 "irrigations" with several unnecessary synthetic fertilizer, insecticide, and herbicide applications for 'hot zone' agriculture guaranteed to barbecue natural lawn inhabitants. Fast-acting inorganic liquids scortch grass and kill vital micro-organisms needed to break down thatch and fight disease. They also bypass the plant and enter runoff, or leave roots without energy to endure summer heat stress. When dead grass fails to decompose, thatch is created, providing a home for fungi and other lawn enemies. Clippings become "yard waste" which cannot be dumped as garbage and requires special - costly - collection.
Yield: Brave New World meets The Killing Fields.

chemicals.

High maintenance regimes are bred to demand attention. Fast-acting, hard-nosed synthetic nitrogen acidifys soil, decreasing pH, reducing worm counts by as much as 2/3 and in general destroying the living pageant of the soil which naturally supplies grass with the nutrition it needs to survive.

With their natural cycles interrupted, grasses beg for nitrogen transfusions to keep them in the green. Herbicides suppress weeds (and, inadvertently, new grass) on dead, compacted soils. Excessive watering trains roots upward, away from soil. Insecticides take aim at bad bugs who come to inhabit thatch which wouldn't be there except that thatch consumers have been eradicated by excessive fertilization.

In desperation, the lawn tender rushes to supply the missing ingredients from the shelves of the grass merchants. The lawn ceases to be a free, active ecosystem and is compelled by a score of symptomatic problems to constant chemical fixes.

High-maintenance lawn shows inevitable effects of chemical burning.

A recent survey of lawn-service culture in a Minneapolis suburb revealed: liquid nitrogen applied on November 2, when it could do absolutely no good other than to feed runoff; an incredible 44 gals. of 13 % nitrogen applied to a 3,500 sq' lawn; broadleaf herbicide applied to kill clover, a plant adapted to supply nitrogen to grasses; a $200 bill to overseed a 600 sq' lawn; phosphorus added to a lawn while in its summer dormant period; 5 applications of 30% nitogen in a single season!

It's a simple fact that lawn owners generally do not know their soil type and its specific nutrient requirements. They rely on advice from retailers who have not tested the soil and whose income results from sale of product. While residential lawns receive little of the pampering golf courses routinely receive - aeration, top dressing, overseeding, topsoil construction - they export 2-3 times the phosphorus into stormwater!

Research shows that nearly 3/4 of all midwestern lawns need no additional phosphorus. Of course, no healthy lawn requires pesticides. Most fertilizers and pesticides migrate off the lawn into the stormwater river. Communities who must restore receiving waters due to polluted runoff may pay as much as $375 per pound of phosphorus removed by detention ponds or diversion.

LOW MAINTENANCE LAWNS

By contrast, a low-maintenance lawn is a partnership with Nature, offering a path to financial savings, ease of maintenance and reduction in potentially dangerous export. While a lawn will always require routine intervention, if well-rehearsed with basic nutrients, a low-maintenance lawn actually prospers from benign neglect, rivaling high-maintenance lawns in thickness without costly synthetic additives.

A low-maintenance lawn presumes that given a chance, every natural actor will play their part, and that of all the vast audience that attends the lawn, only 1% are up to no good. And this small group of hecklers constitutes a large part of

Nutrient-enriched lawn on left; natural lawn on right.

the menu for the other 99% in attendance.

A low-maintenance lawn is high-leafed to shadow soil, slowing evaporation and stiffling weed germination. Carefully selected grass varieties thrive under local climatic conditions. Clippings are left on the lawn, amounting to an annual fertilization. Waterings are infrequent, with deep roots to tide it over in times of drought. Absolutely no pesticides are allowed at anytime! Table scraps (compost) keep it green, plush and disease-free. Aeration every few years deals with compaction, the greatest of all lawn enemies.

Thatch, a fibrous mixture of matted clippings, roots and shoots, disappears in a week thanks to an army of muscular chewers who eagerly remove plant debris. Nor is a low-maintenance lawn a difficult or complex undertaking.

Grass, afterall, is simply the on-stage cast of Earth, a diverse biological mantle of soil, bacteria, fungi, insects, earthworms and tiny microbes. While grass gets the applause, it is the support cast of subsurface characters who break-down, or decompose, plant, animal and inorganic materials into the simple substances desired by grass roots.

Respect the soil and the soil will raise the grass to new heights.

Low-maintenance 'lawn science' starts with the understanding that problems in a lawn generally indicate either: 1) poor seed selection, 2) over/under cultivation (usually too much

EFFECTIVE CARE FOR A LOW-MAINTENANCE LAWN

5 tons balanced soil
100 lbs compost
2 tons air
50 lbs grass clippings
1 reel-type mower

3 - 3½" mowing height
6 lbs selected seed
6.0 -7.0 pH 'sweet soil'
100,000 worms, spiders
less H2O than you might think

Mix soil, air and slow-release organic/compost fertilizer to depth of 12". Add seed appropriate to climate and low-maintenance regime. Mow to 3" height with cleaner-cutting, non-polluting, exercise providing reel-type mower. Add in grass and leaf clippings. Yield: tall, lush growth watched over by a million eyes.

fertilizer, water, mowing, and weed warfare), 3) over-use (compaction), or 4) poor soil construction. Restoring lawn vigor may be just a matter of inviting natural forces to return, or it may require more drastic steps.

Even before weeds arrive, lawn problems may be signaled by chlorosis, a yellowing of grass blades due to nutrient deficiencies, disease or a heavily stressed plant. If plants look sickly, dig out a small wedge of grass (approx. 3-4" square) and measure root length. Roots of less than 4" indicate trouble, usually over-watering or excessive fertilization.

A closer look: above lawn receives weekly commercial treatment. The lawn below is maintained without any chemicals or fertilizers, other than clippings.

Check also for thatch, a brownish build-up of grass parts deeper than 1/2" at the base of the plant due to pesticide poisoning. In a healthy lawn, grass spreads into a thick mat by sending out rhizomes and stolons from which sprout new stems and blades. Blades, when cut, form clippings but do not contribute to thatch unless the many natural plant decomposers are killed off by pesticides. Without them, the tough fiber known as lignin which firms up woody plants, does not break down.

If you can't remove it by ordinary raking, water can't penetrate it either, and a parched, lifeless lawn is a source of polluted runoff. Thatch-related lawn diseases include brown patch, dollar spot and fusarium.

A low-maintenance lawn of course, co-exists with some weeds. In fact, weeds may be nearly invisible though they make up 20% of the lawn! Eliminating all weeds is a futile endeavor: weed seeds are spread by many birds, most often by the common sparrow, and they can remain

RECIPE FOR SOIL TESTING

1 Soil Testing Kit:
Approx. Cost: $10.00
Delivery: 3 weeks
Call 625-3101 for information

SOIL TESTING LABORATORY
135 CROPS RESEARCH - 1903 HENDON AVENUE
UNIVERSITY OF MINNESOTA
ST. PAUL, MN 55108

viable for several decades.

To prevent them from getting a foothold, check the lawn for bare spots and repair. If your lawn is 50% or more weeds, you'd best declare it pileated woodpecker habitat for it needs substantial rebuilding.

To build healthy turf, begin not with the plants but with the soil. Soil problems may linger from home construction as the topsoil, the prime 1 foot of biologically active earth created over 10,000 years of growth and decomposition, is torn away in order to put in a basement. New "topsoil" is likely to be a nutrient-poor mixture of sand and gravel overlain with distant farm-raised sod.

Such imbalances as exist must be addressed beginning with a simple, inexpensive soil test. Only by knowing the character of your soil can you construct a healthy lawn and reduce potential stormwater contamination. If you are the type who wouldn't schedule surgery without first obtaining a reliable diagnosis, by all means determine what your soil needs are before investing in costly, risky treatments.

Healthy soil demands a balanced blend of organic materials and small mineral-bearing rock particles: sand, silt and clay (in descending order of size). Too much of the larger sand and soil won't hold water and nutrients. Too much of the tiny clay particles and water won't drain, producing oxygen-poor, water-logged soil. Blending the proper ratio — about 2/3 sand with the remainder equal parts clay and loam — is the first step to growing a quality lawn.

Soil testing determines the presence or

absence of essential nutrients Nitrogen (N), Phosphorus (P) and Potassium (K). (Keep in mind that many lawn problems result from an excess of N, P or K which burns up the soil's natural community.)

Nitrogen (N) is generally the limiting ingredient in grass production, adding more of it will translate into leaf growth. It is easily washed through the topsoil by over-watering however, and will burn grasses if over-applied.

Phosphorus (P), useful for root growth, comes courtesy of atmospheric fall-out. Phosphorus has the greatest impact on surface waters where it boosts algal growth and should not be applied unless soil testing calls for it. If it does, apply very small amounts of organic phosphorus in the form of fish emulsion, compost, sludge or bloodmeal.

Potassium (K), the third ingredient listed on packaged fertilizer, toughens turf, making it more resistant to drought, disease, climactic extremes and traffic. Minute, but sufficient quantities are generally available in undisturbed, pH neutral (6.5-7.5) soil.

Nutrient recommendations are made in commercial language, that is, so many lbs. of say, 10-5-3 (10% N, 5% P, 3% K) fertilizer per 1,000 sq. feet. The synthetic nutrients offered by commercial products provide immediate green-up ("surge growth") but fail to nourish the biological matrix - the living environment - of the soil. Unfortunately, nitrate, an ***inorganic*** form, and a dominant nitrogen compound in fertilizer - and stormwater runoff - is readily taken up by algae.

The very best soil amenders are ***organic*** and generally do not total more than N+P+K=15. With organic, the medium is the message, as decaying plant tissue (humus) feeds the soil a

RECIPE FOR QUICK SOIL JELLY ANALYSIS

1 qt. jar
1 1/2 c. water
20 shakes

1 spade
5" soil core
12 hours

Dig core and place in large jar. Add water. Shake vigorously, let stand overnight. Measure layers of sand (bottom), silt (middle), and clay (top). Yield: 1" = 20%

well-balanced, complete nutritional package (kitchen compost is a slow-release, comprehensive 2-2-2 fertilizer, 2%N, 2%P and 2% K). Most of the micro- or trace nutrients: sulfur, magnesium, calcium, boron, cobalt, sodium, iron, zinc, copper, molybdenum - necessary to assist the uptake of the major nutrients are present, as are protein and carbohydrates.

Important too, is the fact that organic matter is once-living tissue which provides the necessary texture or 'tilth,' to soil, creating pores for air and water, and presenting microbal life with the building blocks of its existence in an easily digestible form.

Organic matter also neutralizes soil pH, buffering acidic (below 5) or alkaline (above 8) conditions, opening a zone in soil chemistry where nutrients are stripped from soil particles and made available to plants. While organic matter may account for only 5% of soil, it is essential to the state of loam, the perfect integration of soil elements into a home for the millions of unseen creatures upon which your natural lawn depends.

If lost from the yard, tightly bound organic nutrients break down more slowly than inorganic "quick" fertilizers, making less of a splash on the training tables of lake algae.

So take a pro-active lawn stance, leave the clippings and dress it with compost or seaweed (liquid or granular) once a year.

TIP-TOP DRESSING

1/2- 1" Spring Compost
Summer Grass Clippings
1" Shredded Leaves in Fall
1 lb. Nitrogen per 1,000 sq ft

A light Spring dressing of compost will provide a slow-release of nitrogen and minerals in the right quantities without stress. Grass clippings are 90% water, the remainder mostly nitrogen ... leaving them amounts to an annual fertilizer treatment. Covering lawn with thin layer of shredded leaves after first hard frost is another. Or, dress with slow-release solid organic fertilizer: compost, peat moss, manure, bone or blood meal, fish meal (4-1-1), sugar (bacteria food), sea weed (1-1-3), epsom salts (magnesium sulfate), beer (enzyme booster). Apply in late fall.

Look for that "earthy" smell, the result of actinomycetes (threadlike bacteria), a useful microorganism. Nematodes, tiny soil-dwelling roundworms, attack numerous types of grass-chewing grubs. Sowbugs and millipedes feed on dead and dying plant tissue rendering complex proteins into simple nitrogen. Endophytes, naturally occurring fungi, break down thatch and resist insect encroachment.

Worms and their large cousins, nightcrawlers, take leaves and grass clippings from the lawn and drag them underground where they are digested and returned to the surface as worm dung (castings) at the incredible rate of a 1/3 lb. per worm per year! These worm castings are fine humus, equivalent of a 1-1-1 fertilizer, delivered nightly. Earthworms also secrete calcium carbonate, a substance which helps to moderate acidic soil, bringing it up the pH scale to the preferred zone of active biological functioning.

Nightcrawlers burrow several feet into the earth enriching deeper layers of subsoil with organic matter from the lawn while bringing minerals to the surface. Their extensive tunnelling provides for deeper penetration and longer retention of air and water stimulating root growth. After 'hot' fertilizer or pesticide applications however, worms may not return to their chores for an entire season!

Once the natural lawn protectors are in place, you may "proseed" to planting. The following recipes contain grasses which are 'premium' grasses: non-natives bred to grow faster and thicker than natives. (Though they require less water and fertilizer, native grasses clump, leaving spaces for other plants, and lessening their value for curb-to-curb lawns.)

If you don't plan to plant natives, you may still use premium grasses without the fast-

WILD WARM WONDERFUL WORM WAFFLE

1/2 ton leaves
Acid Soil pH increased
Organic Nitrogen, Phosphorus, and Potassium

1/2 ton grass clippings
Alkalinity reduced

Mix in a pesticide-free environment.
Earthworms will consume thatch, aerate soil, reduce compaction, increase drainage and water retention, and mix a ton of nutrient-rich soil for annual yard use. Yield: annual recommended soil aeration and fertilizer treatment.

COOL SEED SALAD (NORTHERN USA)

60 - 80 degrees
summer sleep
perennial ryegrass (England)
Canadian bluegrass

spring/fall growth
white clover
tall fescues (N. Europe)
fine fescues

Mix grass varieties: perennial ryegrasses (fast-growing, hardy & good wear resistance) with tall fescues (drought resistance), Canadian bluegrass (cold tolerance & color), and fine fescues (low nitrogen needs & fine texture). Add a sprinkle of clover (deep roots supply nitrogen to the soil for use by other grasses). Allow to go dormant during high heat of summer. Yields excellent, self-sustaining lawn platform under extreme conditions.

acting fertilizers, the deadly pesticides and the excessive water supposedly required by these hybrids.

The trick is to 'go native' - treat the foreign hybrids with their arrogant needs, as if they were the more humble natives. Purchase a quality seed (there are some 200 'improved' varieties) with a high germination rate (90%) and few weed seeds. Plant at recommended rates and make it adapt to low-maintenance conditions. Assist with clippings, compost top-dressing and occasional aeration. Apply no fertilizer other than clippings; clippings comprise a 1 lb. nitrogen treatment seasonally while low-maintenance recommendations ask for only 1/2 lb.

Cool season grasses (climate zones 1, 3 & 5) grow best north of a line stretching from Richmond, Virginia to San Francisco, California. They are 'cool season' grasses because much of their growth occurs during the moderate tem-

WARM SEED SUCCOTASH (SOUTHERN USA)

85 - 95 degrees
Bermuda grass (Africa)
bahia grass (S. America)

summer growth
zoysia grass (S.E. Asia)
centipede grass

Combine Bermuda grass (quick starting & compacted soil tolerance),zoysia grass (drought & wear tolerance), bahia grass (low nitrogen needs & thatching potential), with centipede grass (fine texture & disease resistance).

peratures of spring and fall. Warm season grasses prefer warmer temperatures, growing throughout hot summer months in southern climates. Of the several hundred varieties available, choose types tailored for your particular lawn requirements: traffic, shade, etc.

Water sparingly, once a week for 2-4 hours or until puddles appear (average sprinkler puts out 1/4" water per hour) if no rain has fallen. (If the lawn is in danger of dying because of drought conditions, call your local government unit and ask them if you may pour water into the dirt.)

If you believe the existing lawn can be spared and only requires limited measures (remediation), you might want to investigate whether your lawn has fallen prey to the most common of all turf-grass maladies: compaction. Heavy traffic — kids, the mailman or paperboy, the path to the car — affects the grass by affecting the soil. As the soil is compressed, roots can't penetrate and remain shallow, unable to tap water and nutrients. Worms have a difficult time reaching the surface so necessary aeration is halted. Essential microbes and bacteria are denied living space promoting the accumulation of thatch. Don't be slow to aerate, this could be the most important step you can take toward a healthy lawn.

WATER TIGHT DELIGHT

1" total per week,minus rain, til mid July
1/2" every other week til Fall if in drought

Do not over-water; over-watering brings roots to surface and results in thin lawns, compaction, weeds, fungus, thatch and destructive insects. Water early in the day - an occasional soaking is better than a light, daily sprinkle. Cool season grasses go into a state of "plant rest" until fall. Gradually reduce watering to condition lawn, limiting sprinkling to 1/2" every few weeks under conditions of mild drought. Limit heavy traffic.

RECIPE FOR MOW-A-TERRAE-YUM

2-2 1/2" height spring
3-3 1/2" height summer
2-2 1/2" height fall
1 sharp blade

6" better root depth
50% greater leaf surface
8,000 weed shields per foot

Mix mowing, a violent shock to the roots, with periods of taller growth, to build roots. Add increased leaf size for greater photosynthesis. Cut 1" off when grass is 4" tall. Yield: Moist lawn, sufficient shade to prevent hot-weather emergence of dandelions and crabgrass.

Improved lightweight reel-type mower on the right, if kept sharp, offers similar ease of cutting at lower cost and with greater environmental — not to mention exercise — benefits than the model on the left. Never aim clippings toward the street! Sweep up any clippings on street, driveway or sidewalk and compost.

Red sticky trap, although indiscriminate, protects fruit from apple maggot fly foe without polluting stormwater.

6
A CIRCLE OF POISON

Whatever combination of lawn, theme gardens, natural environments and more traditional annual, perennial or vegetable gardens chosen, you will want to weigh very carefully any decision to douse your real estate with poisons. Today's toxics are more potent and widespread than ever. They exist in the unique 20th century delusion that mass murder makes sense when directed at the backyard. Of course, virtually every sample of stormwater shows deadly chemical traces.

"Pesticide" comes from joining the Latin words *pestis* ("plague") and *cidere* ("to kill"), to form a broad category of substances capable of exterminating what plagues us. Pesticides may be *biological* - sometimes called "botanicals" - derived from natural plant and animal material, or even from bugs themselves. Most (some 65,000) are *synthetic*, that is, man-made.

Synthetics are derived from petroleum and aim for the bug's central nervous system. Synthetics are generally more potent than natural pest inhibitors, and they persist in the environment longer being more immune to the forces of sun, rain and digestion. Synthetic pesticide residues are more likely to threaten surface water than are botanical controls.

A pesticide aimed at plants is a herbicide. Targeted at bugs, it's an insecticide. Intended for rats and mice, it's a rodenticide. Formulated to deal with 'lower' plants - molds, rusts, mildews, and bacteria - they are fungicides.

The first synthetic pesticide, DDT, was developed during WWII and was banned in 1972. A persistent deadly pesticide, DDT was working its way up the food chain when Rachel Carson pulled the plug with her landmark book, "Silent Spring," published in 1962. While "Silent Spring" slowed the growth of persistent or 'hard' insecticides, less long-lasting but more toxic herbicides have increased threefold.

RECIPE FOR DEADLY HARVEST

methyl parathion
MCPP (Mecoprop)
DDT (RIP* 1972)
aldrin
atrazine (RIP 2000)
toxaphene (RIP 1983)
diazinon (Spectracide) (RIP 1990 on golf courses)
dieldrin (RIP 1983)
glyphosate "Roundup/Kleerup"
sevin "Carbaryl"
2, 4, 5-T (Agent Orange)
EDB (RIP 1983)
4 million lbs. dioxin
cyanazine "Bladex" (RIP 2000)
chlordane (RIP 1988)
chlorpyrifos (Dursban)
2, 4-D (Weed-B-Gon) (Trimec)
heptachlor (RIP 1983)

Mix indiscriminately. Yield: In humans, increased risk of breast cancer; central nervous system damage, lymphoma, infertility, prostrate & testicular cancer and deformation of male sex organs; in birds, seizures, deformed bills; tumors in pets; fish kills.

Pesticide use has become accepted practice despite alarming signs. Today pesticides are found in many commercial products including paint, lumber, detergents, carpeting, shower curtains, furniture, flea collars, shelf paper, etc.

The 18,000 golf courses in the U.S. apply pesticides at rates up to 20 lbs. per acre. Some golf clubs report increased cancer rates among members who place golf balls and tees in their mouth.

Lawncare companies service 7 million acres of lawns, turning them into "chemical landscapes." Homeowners add nearly 10 lbs. of pesticides per acre. Three out of four Twin Cities lawns are treated with pesticides although tests in neighboring Wisconsin demonstrate that only one lawn in 200 actually needed pesticides. Low-maintenance lawns almost never need them. Suburban communities fear not spraying herbicides in public parks because of the 'dandelion backlash'!

A few pesticides, like chlorine bleach, a bacterial exterminator, are so commonly used in hospitals, public facilities, even the home washing machine, that we tend to forget that all pesticides are poisons. Pesticides are labeled by the Environmental Protection Agency (EPA) according to their **acute** toxicity (causing harm with a single dose, as oppossed to **chronic**, or repeated exposure): CAUTION (slightly toxic); WARNING (moderately toxic); DANGER (may not kill but can cause severe, permanent injury); and DANGER--POISON (highly toxic, a few drops can kill).

In 1984, a leak from a pesticide plant in Bhopal, India, killed 2,000 people and injured 150,000.

A billion pounds of pesticides are released annually in the U.S., 65 million pounds in Minnesota, with urban application rates steadily rising. Virtually every man, woman and child in the U.S. is exposed on a regular basis to pesticides. Numerous pesticides have been banned because their long-term residual effect has been found to be carcinogenic or mutagenic (causing biological mutation) in humans.

Some researchers believe that man-made pesticides may be scrambling genetic instructions leading to a feminization of the planet, with smaller male genitalia and impotence appearing in several species.

Food crop loss has actually increased since synthetic pesticides were introduced 50 years ago indicating that insects may be building immunity. Sadly, it is children with their lower thresholds for dangerous accumulations that have the most to "gain" from the increase in synthetic pesiticides.

Because they eat more fruits and vegetables than adults, children may have already received *lifetime acceptable doses* of commonly used pesticides by the time they are in kindergarten. Much of the exposure comes from airborne

Species diversity
Botanicals
Beneficial Parasites, Predators

Companion Planting
Insecticidal Soaps

Add ingredients to reduce Pesticide use by 90%. Serves everyone.

dust blown long distances.

Misuse of pesticides, 'if it crawls, it falls,' knock-down applications kill virtually everything within a chemically-rich circle of poison. Beneficials - pollinators, useful predators - are blasted creating the run-down housing that attracts a lower class.

Diazinon, a powerful insecticide highly toxic to fish and wildlife is, thanks mostly to homeowners, found in virtually every sample of runoff reaching local receiving waters. Diazinon is said to cause the deaths of several million songbirds annually in the U.S.

Herbicides commonly used for dandelion control on urban grasslands, are measureable in nearly all stormwater samples.

Three-quarters of the pesticides banned for use in the U.S. are manufactured and used in the Third World where they come back to us either on imported fruits and vegetables or as dust falling from the atmosphere.

Research suggests that pesticide synergies - pesticides acting in combination - may have 1,500 times the potency of individual effects!

A saner approach - Integrated Pest Management (IPM) - uses simple, common sense controls to battle unwanted creatures without toxifying the world.

IPM's **'smart'** stragedy employs: a) ***cultural*** factors (what people can do to deny pests food and lodging); b) ***biological*** life (good insects to manage bad insects); and c) ***natural*** chemicals (low-impact, plant derived toxics), to reduce, but not eliminate pests.

To eliminate an outbreak of pests in an agricultural setting in order to save a crop makes sense, but in an urban setting, on a global scale?

RECIPE FOR BOTANICAL ATTACK

NEEM (Asian Neem Tree)
sabadilla (Mexican lily family)
pyrethrum (chrysanthemum plant)
rotenone (tropical pea family)

Mix the plant extract NEEM with pyrethrum for powerful punch lethal to numerous insects, including ants, aphids, cutworms, striped cucumber beetle, roaches, spiders, centipedes, wasps and hornets. Rotenone and sabidilla kill Colorado potato beetle, cabbage worms, squash, chinch and stink bugs. Serves to lessen damage on humans; reduce impact on upland and aquatic wildlife. Spray carefully!

CREEPING CHARLIE CRASHAROLE

5 t. 20 Mule Team Borax
1 qt. H_2O
25 square-foot yard area

Mix in a sprayer. Use on invasive creeper -- sold by some garden centers for its agreeable odor.

Nature, alas, confounds us. What we may regard as a pest is, in fact, a formidable opponent, designed over a history 1,000 times longer than our own, to play a complex role in the life of the planet. By going for the knock-out punch on a single species, we unconsciously flail away at beneficial insects, birds, butterflies, worms, fish and a host of other innocents woven together in a necessary tapestry of mutual dependence. The result of our meddling often is a barren battlefield which may take years to recover.

IPM strives initially to make plants healthy. Healthy plants survive low-level attacks, whether it be mild drought, malnutrition or hungry insects. If the attack intensifies, IPM seeks to remove the staging area for the onslaught. Stronger measures include the marshalling of an army of predatory destroyers for a counter-attack, and, finally, as a last resort, the use of natural chemicals for quick, specific, temporary toxic knock-down.

Assuming you've followed steps in Chapter 5 for creating healthy soil, IPM next selects plants and places them to take advantage of known synergies to provide a valuable wall of self-defense. This technique is called "companion planting" and serves to increase available nutrients, alter pH and deter pests.

Companion planting pairs different plants which have been observed to help each other, even when the precise mechanism of assistance is unknown. Examples include planting an occasional radish among cucumbers because they repel cucumber beetles. Similarly, nasturtiums discourage ants, and thus aphids, since ants often herd aphids as livestock for their sweet secretions.

Other **cultural** practices include: plant rotation, so specific plant pests don't become too comfortable; mulching to deny weed seeds the light they need to germinate; diversification of plants so as not to set the table too abundantly for specific pests; inspection to insure that curling, spotting and yellowing leaves do not go unattended; identification of lawn or garden maladies to determine the specific pest responsible and suitable low-impact response; and early intervention to limit plant damage—and treatments—to undetectable levels.

RECIPE FOR GUIDED MISSILE PLANTINGS

nasturtiums (repel aphids)
marigolds (repel Mexican bean beetles)
carrots (protect tomatoes)
tansy (striped cucumber beetle)
catnip (beetles)
daisies (all)
onion, thyme, hyssop (protect broccoli, cauliflower or cabbage)

basil, with asparagus (flies & mosquitoes)
radish (cucumber beetles)
garlic (protects roses from black spot)
mint/rosemary (cabbbage worm)
zinnias (mosquitoes!)
geranium & chrysanthemum (most insects)

Mix for natural, low-impact protection from nuisance insects.

Established pests can be controlled through **biological** methods, that is, creature vs. creature. Every pest is prey to an insect/avian/ mammalian predator.

Mosquitoes, for example, are difficult to control in the larval stage because they mature in a pinhole of stagnant water; as adults they fly 10 miles a day in search of 'precious blood.' A wet spring will hatch billions of these irritable vampires, fomenting in a single suburban beer can more indiscriminate violence than all of television.

Bats, however, are nature's premier mosquito controllers, consuming 1,000 an evening all summer long. Land stewards will spurn indiscriminate pesticide foggers and bug zappers for this furry smart-bomb. Bats nest near water and tucking a bat house under the eaves may be the last investment you will have to make in mosquito control.

The gentle-looking *green lacewing* and the smallish *lady beetle*, have giant appetites for aphids, one of those potentially explosive populations that must be kept in check. Tiny *brassicae* wasps make excellent bio-controls lethally parasitizing common garden pests - cabbage worms, loopers, corn earworms - and can be ordered

RECIPE FOR GOOD BUG CHOW

angelica
eunonymous
yarrow
goldenrod
marigold
butterfly weed
tansy
queen anne's lace
dill
cosmos
coreopsis
alyssum
chamomile

Plant ingredients to brew nectar & pollen sources for beneficial predators. For an insect snack, mix 4 parts sugar to one part water, add 1 cup brewer's yeast.

RECIPE FOR PREYING

200 ground beetles (carabidae)
10,000 spiders
1,000 parasitic wasps
aphidildae, trichogramma, braconid, & chalcid wasps
500 green lacewings
1,000 lady bugs
500 assassin bugs
500 damsel bugs

Put ground beetles, large black tanks trimmed in blue, green or purple, on patrol against slugs, snails and cutworms. Lacewings and Lady bugs, the "aphid wolf," can eat 100 aphids per day while an army of spiders in lawn and garden ensnare thousands of flies and mosquitoes every hour. Apply parasitic wasps to control scale, whitefly, cabbage worm, codling moth and web worm. Larger animals – tomato hornworms, cabbage loopers, stink bugs – can be hand-picked.

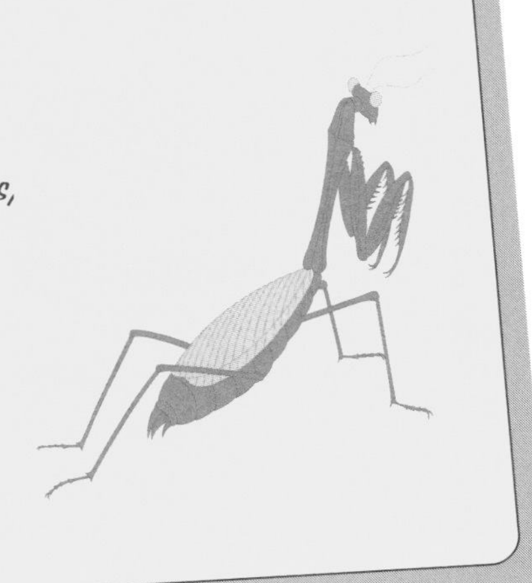

through the mail.

Such predator shopping can do the job the natural way protecting surface runoff.

If, in order to avoid another Alamo, you're willing to risk concentrated fire, extracts of certain plants can be distilled and combined with alkali (mineral salts) to produce **natural chemical** pesticides. Derived of once-living ingredients, they break up sooner and pose less risk of leaking into the stormwater river.

Pyrethrum, rotenone (used for centuries as a fish poison), Bt (*bacillis thuringiensis*), rubbing alcohol, insecticidal soaps and oils are examples of kill-all 'natural' chemicals. Even NEEM, the growth-inhibiting miracle natural from India, cuts a wide swath among the insect haves of the world.

Pheromone lures to trap and kill mature males can be tailored for individual species and are

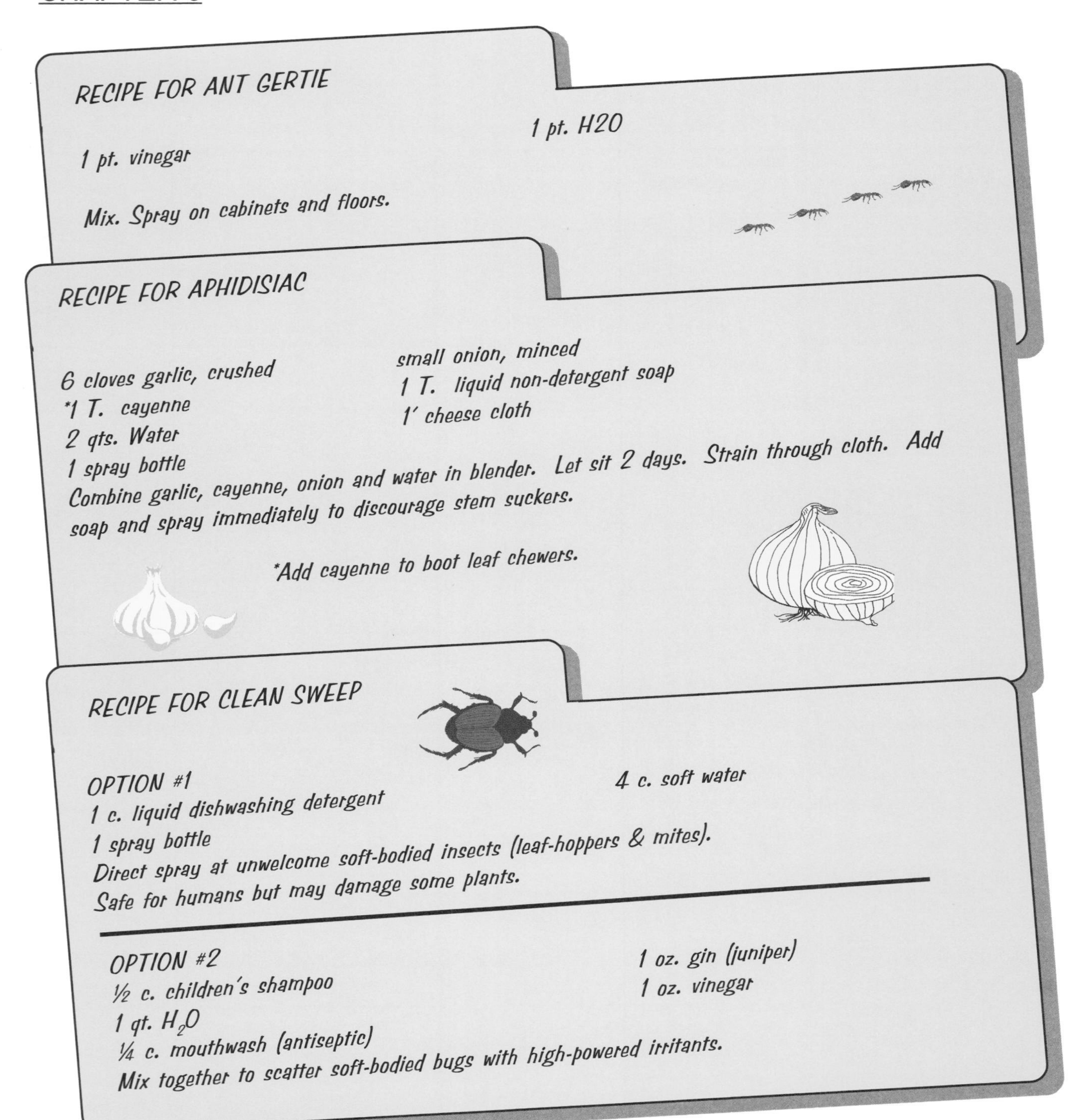

probably the safest of the *naturals*.

Bio-controls are more time-consuming and more expensive than synthetic poisons but they are more environmentally friendly.

Mulching is what pesticide dreams are made of. Good mulching suppresses weeds while nurishing the soil and protecting runoff. Mulch is a soil covering that discourages 'volunteers' in garden and landscape plantings by shielding the soil surface. Weed seeds that don't get sun, don't germinate; and weeds that do manage to sprout, die when they can't root.

Summer mulches cool soil, keep it manageable and improve root development of desirable plants. A layer of mulch during the growing season prevents soil-borne diseases from reaching and infecting plants.

Winter mulches trap air around the plant moderating the freeze-thaw temperature extremes which explode plant tissue damaging the plant.

Minneapolis Rose Garden under 2' winter mulch.

Mulches made up of dead organic matter — leaves, bark, wood chips, cocoa hulls, etc. — are pervious, allowing rainwater to get to plants. They also replenish the soil as they breakdown and decompose. But mulches made of the sons of imperviousness - rocks, landscape fabric, plastic - do not.

RECIPE FOR MULCH MADNESS

Sunday ad section
10 bushels chipped wood/bark
3-4 bushels compost
10 bags fallen leaves
5 bags evergreen needles

Mix for natural, low-impact protection from nuisance weeds.

Spread overlapping newspaper sheets 5 deep under and around plants, shrubs and trees. Shred leaves to prevent matting. Mix leaves, needles, bark and compost at a 1 to 1 ratio. Cover newspaper with mixture to a depth of 4". Every year, add 2-3" leaf mixture to maintain. Mulch from within 4-6" of the trunk outward to drip line, i.e., the reach of the limbs. Yield: Mulching saves water, eliminates chemicals, makes for healthy trees and cleans stormwater.

'You can take the American out of the car,
but you can't take the car out of America.'

7

CARSICKNESS - THE END OF AUTOCRACY?

The family automobile, along with homeownership, is symbolic of American economic success. In the U.S. at least, democracy, an open press and secret ballot are no less important than the dream of owning a fully-restored 1958 Pontiac *StarChief*. Unfortunately, its universality makes the car a major threat to water quality. Research shows a direct link between daily mileage and water quality with the most heavily travelled roads producing the most polluted runoff. Some 200,000 vehicles chose to die in Minnesota each year. Their end couldn't be more Kervokian.

Even a well-maintained automobile is one ton of shedding, leaky animal. Parenting any creature this size requires patience, love, money and a ready supply of hand towels. As this valued pet ages, it spawns 30-some bodily wastes - including poisonous fluids and several toxic metals: aluminum, arsenic, cadmium, chromium, copper, lead, mercury, nickel, and zinc. All can have catastrophic effects on lake inhabitants.

In addition, Minneapolis applies about 20,000 tons of salt (sodium chloride) annually to major

RECIPE FOR AUTO-MATING LAKES & STREAMS

1,800,000 all-weather personal conveyances
500,000 detergent applications
2,000,000 gal. old motor oil
10 tons antifreeze, 14 winters
30,000 mi. pavement
100 tons salt
150 tons exhaust

Add ingredients in multi-county area. Mix in air and stormwater river. Yield: acidic snow, fog and rain. Serves to erode concrete and steel into toxic tide destroying native plants, wildlife and aquatic species.

streets, and a similar amount of sand in residential areas. Salt is used to melt ice; sand increases traction in order to allow autos to speed at a rate they're used to. The victims are aquatic vegetation, invertebrates and the auto itself: 50 percent of automobile corrosion is due to chloride in road salt.

If you are mature enough to own a car, you are mature enough to clean up after it. Allowing your auto to soil lakes and streams is unacceptable behavior. In most states it is illegal to abandon cars or to dump auto wastes: gasoline, motor oil, antifreeze, ashtrays, etc.

Unfortunately, for reasons no doubt linked to serial murder, some car owners abuse

RECIPE FOR COLORFUL CORROSION CASSEROLE

4 gal. enamel paint
30 lbs. (old) bumpers
1 gal. leaks
5 lbs.cancer-causing hydrocarbons
15 lbs. nickel trim
30 light switches
nitrous oxides
5 lbs. Bondo

Pour when no one is looking. Stir into air and water. Fill street until 'rainbow' hued. Yield: acid rain; ear, eye, mouth and throat infections.

Don't let this

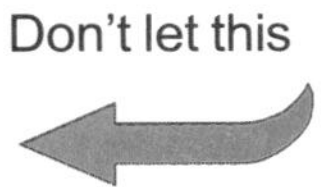

Become this.

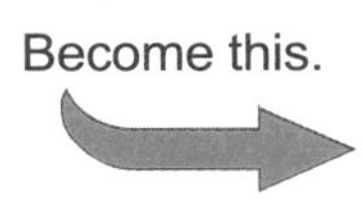

Instead of sand in plastic bags, use weights or bricks for winter traction. If you must carry sand, put it in a 5-gallon bucket with a secure lid.

their vehicles and the streets which harbor them. These tragic individuals are guilty of *automalpractice* and, in the best of all-possible worlds, would be sent away to 're-education' camps for an unspecified period of time. Since we lack the freedom to be spared such idiots, we must close ranks and ridicule them into submission. When you spot these fools pouring motor oil into the street — 200 million gallons are illegally dumped each year — or, emptying ashtrays, fast-food bags or mats in the path of stormwater, remind them that it's on its way to what used to be the best beach in town.

Nearly all of the automobile, including its fluids, can be broken down into easy recycling options. Retail auto parts stores, service stations and municipal or county 'auto recovery' collection centers can accommodate the altered states of our non-ferrous family friend.

When changing fluids, use a drip pan and pour old materials back into newer container for recycling. Don't mix discarded fluids.

These fluids may be recycled or collected locally: antifreeze, automatic transmission fluid, brake fluid, diesel fuel, gasoline, kerosine, wax and polishes with solvents, and motor oil. These substances also have in common a lethal power and must not be released to streams or lakes via

WINDSHIELD, HEADLIGHTS & CHROME CLEANER

1/2 c. baking soda
1/2 c. vinegar
1 gal. warm water

Apply with sponge wearing rubber gloves. Let dry, then wipe.

RECIPE FOR LOW-LAKE IMPACT CAR WASH

1 lawn or gravel driveway
commercial car wash
dozen absorbent pads
box non-phosphate soap
can biodegradeable wax
1 minute irrigation, rinse

Wash at commercial locations connected to wastewater treatment plant. Use non-phosphate cleaners. Or, park on lawn for absorption of soapy water, auto wax and underbody grit. If you use a street parking pad while washing, limit rinse to 1 minute. Place absorbent pads around vehicle to reduce pollutant flow. Move charity car washes from parking lots to soccer fields.

RECIPE FOR ANGRY ANTIFREEZE ANTIPASTO

3 oz. antifreeze
several pets
million gallon drinking water
1 adult, deceased
1 child, deceased
billion microorganisms

Swallowing even small amounts of antifreeze can kill an adult, child or pet. Antifreeze wipes out necessary microorganisms in septic systems, sewage treatment plants and lakes. Recycle to service station or repair shop. Do not mix with used motor oil, pour into street, alley, ground or down sinks and toilets.

AUTO DEGREASER (FLOORS, FLOOR MATS)

1/2 c. corn meal
1/4 c. baking soda
1/4 c. salt

Mix according to size of grease/oil puddle or stain. Rub in gently and allow spot to be absorbed (2-4 hours). The residue, whether vacuumed or swept is, strictly speaking, hazardous waste and should be 'solar cured' (2-3 days thorough drying), then transported to appropriate county drop-off site.

stormwater. Ask your friends and neighbors for their contributions until you have a full trunk load. Tell them it's their turn next. It is illegal in Minnesota to dispose of used motor oil or used oil filters in the trash or anywhere else on land or water.

Spotting under your car indicates that the engine, transmission or radiator is leaking and needs attention. Fixing the leak quickly means saving cash, the car and water quality.

Simply keeping tires properly inflated amounts to a 5% fuel savings, while periodic

RECIPE FOR MATURE MOTOR OIL

5 qts used motor oil
drip pan
1-800-RECYCLE (in Mpls 673-2917)
large plastic jug
service station

Pour oil into plastic jug, label and return to service station or government-approved motor oil collection center for recycling. Do not incinerate, mix with trash or spread as non-specific herbicide. When burned, motor oil creates hazardous aerosols; landfilled it poisons ground water; poured on soil to contain dust or kill weeds, it renders vast areas sterile and may leach to surface water, ruining drinking supplies and infecting swimmers.

CAUSTIC CAR COCKTAIL

8 oz. transmission fluid
gal. windshield cleaner
16 oz. polish
12 oz. hub brightener
10 oz. brake fluid
slug a' grease
qt. upholstery cleaner

Buy wisely, apply sparingly, dispose of properly. Don't mix. Donate unused flammable, toxic ingredients to local auto-repair class. Car interiors should never be "cleaned" into streets. Vacuum, broom, or shake out floor mats over yard. Absorbent mats should be used when changing fluids to absorb spills. Treat soiled mats as hazardous waste; dispose of at county collection sites.

tune-ups earns a 20% mileage increase. That's a significant reduction both in the cost of driving and in the charge to surface waters in hydrocarbon and nitrous oxide residues. Purchase a tire gauge and use it at least once a month (tires lose a pound of pressure - 1% of mileage - for every 10-degree drop in temperature); have the car tuned every 10,000-20,000 miles. Remember, cars getting less than 22.5 miles per gallon are official "guzzlers."

BASIC BATTERY BAKE

1 gal. sulfuric acid
1 battery casing
18 lbs. lead
1 resistant, leak-proof container

Combine dangerous lead and acid with battery casing. Convey to retail outlet or alternative in leak-proof container. Many states mandate including recycling cost in the purchase price. Yield: less fish killed.

CARTOON CAPER

new thermostat
3-4 clamps/new hoses
front wheel alignment
tire gauge
spark plugs

Mix proper tire inflation (a 5% gas mileage saving), with a tuned engine (20% mileage increase), wheel alignment (2% mileage savings) & thermostat (7%) for major reduction in polluting residues in the air, on streets and in the water. Good hoses prevent air conditioner and antifreeze leaks.

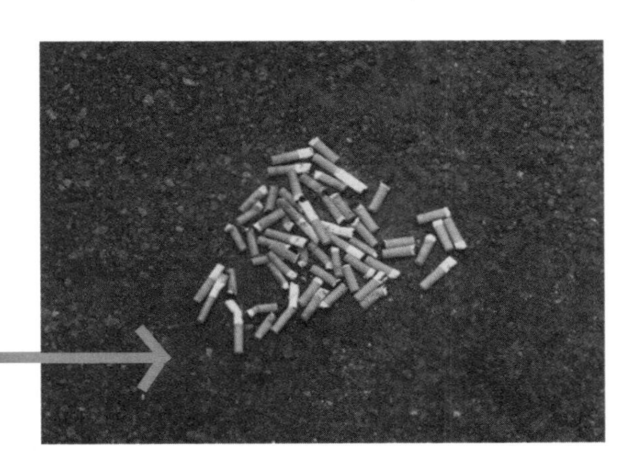

The auto: a stormwater debris delivery vehicle?

RECIPE FOR ALTERNATIVE TRAVEL

25 mi. mass transit
3 mi. walks
50,000 car poolings
100,000 bicycles

8

PET PEEVES - SEE SPOT SPOT

10 MANDATORY MUNICIPAL CODE

CHAPTER 64. DOGS AND CATS

64.50. Leashing; feces clean up.

(b) Any person having the custody or control of any dog or domestic animal shall have the responsibility for cleaning up any feces of the animal and disposing of such feces in a sanitary manner. It shall furthermore be the duty of any person having custody or control of any dog or domestic animal on or about any public place to have in such person's possession suitable equipment for the picking up, removal and sanitary disposal of animal feces. Conviction shall be punished with a fine of $100.

Infamous movie director, John Waters, climaxed the pinkest of all American screen satires by having his leading unlady, the 300-lb. *Divine*, shovel dog poop into her mouth. Though it sounds bad — and looks even worse — Waters' cinematic shocker was not only a side-splitting exploration of bad taste, it was also, unfortunately, documentary urban reality.

Normally, few persons will discuss the composition of animal feces, unless of course, you're eating. Not a pretty subject, the contamination of lake, creek and river water by animal wastes. Still, an examination of fecal content in stormwater may explain some mysterious and serious illnesses.

Potential impacts include *parasites, bacteria* and *virus*es producing painful, even life-threatening reactions in human hosts — and fecal solids influencing lake and river biology with heavy loads of *phosphorus* and *nitrogen.*

The sad truth is animal feces are a *"significant source"* of urban stormwater pollution. Once in water, pet feces reduce dissolved oxygen, release toxic ammonia and add nutrients. Those of us who venture into local waters are swimming with — and consuming — fecal matter from any number of wild and domestic animals.

Nationally, there are several thousand beach closings each year due to fecal stormwater contamination. According to an informal convention among states, beaches are inspected for

RECIPE FOR BOW WOW POO POO

1 c. bacon fat
2 T. yam flour
1 t. sodium tripolyphosphate
4 c. dried animal digest
roundworm parasite
1/2 c. soy hulls
3 T. cellulose gum
2 c. pregelitanized dried corn meal
crypto sporidia
3 t. (each) yellow #5, red #40, blue #1

Mix ingredients in large muscled chamber with gallon canine digestive juice until brown and thickened. Squeeze onto curb. Yield: swimming companion.

indicator bacteria twice a month, or when an illness has occurred. This 'convention' does not have the force of law.

High fecal coliform numbers (over 200 per liter) in several water samples, indicate a general contamination by feces and the presence of serious bacteria. Fecal coliform counts after rain events are often 10 to 100 times this standard. Almost all stormwater contains bacterial counts in the range of raw sewage.

Water of course is an excellent medium for the exchange of bacteria, 80% of all disease transmission is water-related. Most such illnesses are intestinal leading to acute abdominal pain, cramping and, oh yes, diarrhea.

Escherichia coli, vibrio cholerae, *salmonella* and *shigella* are bacterial infections transmitted to humans by other animals. *E. coli gastroenteritis* is a leading cause of infant mortality worldwide. Surface runoff carries this bacteria to ground water drinking sources, or to recreational waters. Children are most at risk because they ingest more water while swimming and lack adult immunities.

Cholera bacteria are among the most deadly with past epidemics killing thousands. Outbreaks are rare in this country, and usually traced to contaminated drinking water.

Salmonellosis is the most common cause of bacterial diarrhea. Symptoms include fever, muscle aches, headache, vomiting, and diarrhea. One type of *salmonella* causes typhoid fever.

Shigellosis, also known as dysentery, has symptoms similar to *salmonellosis*, with dehydration a cause for concern in children.

Campylobacteriosis is another bacterial infection carried from dogs and cats that frequently causes diarrhea in humans.

Cryptosporidiosis the potent intestinal parasite that laid low thousands of Milwaukeans in 1995, can be sourced to cows, dogs, even raccoons. ***Crypto*** can last months in an aquatic environment and its symptomatic diarrhea is difficult to treat, sometimes resulting in death to individuals with weakened immune systems.

Toxocariasis, or roundworm, can be transferred from dogs to humans by touch. May cause cough, fever, rash and vision loss.

Toxoplasmosis is carried by cats and can cross the placental barrier causing birth defects in fetuses of pregnant women. (Anyone working the soil in gardens where cats defecate should wear gloves and peel all vegetables.)

Most communities do not test thoroughly enough to detect these threats. Since the tests only have to be conducted every two weeks, it's

RECIPE FOR DON'T DO-DO LOAF

2 paper towels
1 plastic pooper scooper
puppy love
1 large plastic sandwich bag
1 toilet flush

Grab wastes with paper towel or scooper, put in sandwich bag. A plastic sandwich bag is thicker and easier to work with than a plastic newspaper sleeve. Transfer to toilet and flush. Wash bag in hot, soapy water weekly. Yield: One cool stool.

likely that last week's bad bacteria have shriveled and died under a relentless summer sun.

Smart swimmers know that rain events 1/2 inch or more spawn stormwater and potential bacterial outbreaks from animal feces. Health Department personnel suggest not swimming in surface waters near stormwater outfalls within 2 days (48 hours) of a rain event.

Let's face it, but not eat it: fido's poop is a lot like human poop, say veterinarians, being mostly cornmeal, fats and protein. Parks speculates that 35,000 annual dog "walkings" around the lakes are for the specific purpose of dropping the equivalent of 5 dirty baby diapers into a link in the Chain of Lakes. No one has ever heard a really good excuse for not cleaning up Spot's

KITTY KORNER

1/2 lb. beef tallow
1 lb. poultry giblets
1/3 t. titanim dioxide
litter box
3 gal. veal, pork, lamb, dried animal digest
1/2 c. carrageenan
1/4 t. vegetable gum
2 c. fish by-products
20' chicken wire

Use chicken wire to keep cats out of gardens and litter boxes to keep cat scat out of lakes. Remember that even a well-fed, declawed cat – if not confined – is a skilled predator of birds and small mammals. A single free-roaming cat can kill as many as 1,000 animals annually; 30-40 million Minnesota song birds are killed by cats each year.

spots.

Pet owners who 'curb' their animals i.e., allowing it to defecate into streets or on boulevards, without pick-up, are polluters. The depositing of pet feces on public property without prompt removal is specifically forbidden and punishable by fine.

Pets are pets because they remain perpetual children. No one should seek to own a child who's unwilling to take care of its basic needs. Road rage is but a scratch compared to a bather's anger at the sight of one of these huge breeds uncoiling a 1/4 pounder within feet of the water while its owner studies a cloudless sky for hint of rain. Then the two of them hustle off in a gay mood, as if their little larceny has disappeared in a puff of non-polluting smoke!

Anyone, whether it's blocks from the lake or on its thin apron of turf grasses, must pick up pet wastes in this easy 3-step method:

1) All animal poop must be picked up and flushed down the toilet to insure proper treatment.

2) Burial in a hole 6" deep is a secondary removal option.

3) Placing in garbage is a poor alternative, while wrapping in plastic and placing behind a tree is totally unacceptable.

Nor should geese be tolerated shoreside. Geese graze grass grandly, the feces from a single bird contributing the necessary phosphorus for 25-30 lbs of algae. These noble beggars do not belong where they can contaminate swimming beaches or downstream flows. They would make excellent Thanksgiving Day banquet guests at city charities.

1997 *Southwest Journal*:

"On April 8, police observed a car parked along East Lake of the Isles Parkway with its emergency flashers on. They also observed an unleashed English Springer Spaniel defecating in the park near the vehicle. The driver of the car, a 28-year-old East Calhoun man, then drove about a half-block as the dog followed. The man called the dog, which then jumped into the suspect's car. Police issued the man a citation for not cleaning up the dog feces and for not leashing his dog."

9
WHY WETLANDS?

The concensus among scientists is that watershed land use - and the **type, number** and **placement of wetlands** - has a pivotal impact on lake water quality within a specific drainage area. In fact, one can just about predict that in cities of 100,000 or more, impervious surfaces (streets, sidewalks, roofs) approach 50% of land use resulting in eutrophic, highly fertile conditions downstream <u>if no wetland treatment of runoff is available.</u>

Street-assisted urban runoff detaches nutrients (phosphorus, nitrogen), sediments (soils), bacteria, and heavy metals (mercury, cadmium, lead, zinc) and exports them to low-lying basins during rain events. Failure to trap this slurry and remove its concentrated pollutant load dooms urban water to rapid deterioration.

In areas where land cover is continually disturbed by human activity, and public works and education inadequate, wetlands are essential to treat the flow of pollution-laden runoff from ungoverned streets and alleyways.

In a natural state, surface waters - streams, rivers and lakes - are attended by wetland fringes because they could not survive otherwise. Wetlands aren't just future golf course or pastureland. Rather, they are living scaffolds for the rejuvenation and support of open water.

A "natural" water purification system.

CHAPTER 9

To achieve a stable existence, a human family requires social skills, employment, a network of friends, housing, savings, etc. Lakes and rivers too, require certain conditions for their continuence.

Open waters maintain a balanced ecosystem because they possess means to deal with the tide of upland debris stirred to life by lightning, fire, erosion, seasonal flushing, and development. They are protected from major fluctuations in water levels which can submerge or uproot habitat niches denying inhabitants reliable sources of food, shelter and reproductive privacy. Nor do they worry about huge pulses of walleye-frying acids or algae-fying nutrients. Desirable recreational waters depend on wetlands to act as an early line of defense.

At one time, wetlands comprised almost a third of the continental United States. If more than half of that original wetland acreage had not been drained, our climate would be more moderate, there would have been no disastrous floods in 1995 and 1997 and several extinct species including the passenger pigeon would still be with us. (Continued drainage at today's rate of 100,000 wetland acres annually, now threatens the $25 billion coastal fishery industry.) Wetlands have proved to be so valuable to inland waters, particularly urban waters, that they are being rebuilt as the cheapest, best means of dealing with urban runoff.

Whether natural or "engineered," wetlands are complex biological systems which serve to cleanse water passing through them. They do this by performing several amazing feats which trap, transform and absorb stormwater pollutants.

While often lumped together as "swamp," wetlands include elements of true aquatic systems, terrestrial systems, and intermediate systems. Wetlands, of course, are "wet" — low areas which collect and hold enough water to create unique plant and animal communities. Seasonal or continuous saturation builds a transition zone between crumbly, well-drained soils called 'uplands,' and the open water habitats of lake and stream.

Because water levels rise and fall regularly, wetlands host an array of differing depths and lengths of saturation, creating micro-environments shared by species of both aquatic and terrestial realms. Over 1,000 species of plants and animals owe their lives to wetlands making them second only to rainforests in productivity.

Wetland types cover a spectrum from "occasionally flooded" <u>flats</u>, to <u>wet meadows</u> (broad-leaved plants, grasses, sedges), <u>shallow</u>

marshes (cattails, arrowheads, pickerelweed), deep marshes (wildrice, pondweeds, waterlillies), shallow, open ponds (less than 10' deep), shrub swamps (alders, willows near sluggish streams), wooded swamps (tamarack, spruce, balsam), and bogs (sphagnum moss, cranberry, shrubs) - with subtle differences playing important roles in the lives of many creatures.

What wetlands have in common is sufficient water at some point in the year to 'drown' the roots of terrestrial plants, favoring aquatic plants, owners of special straw-like stems that enable them to grow in oxygen-starved, waterlogged soils.

The lack of oxygen in wetland soils means organic matter does not undergo breakdown and decomposition by bacteria and soil microorganisms. Nor is it likely to be burned in the vast, disastrous fires that take a huge toll in lives lost, resources wasted and climate-altering carbon dioxide build-up. Instead, wetlands accumulate organic material in rich layers eventually turning it into peat and other fuels. These processes take thousands, even millions of years.

Yet, while wetland soils lack oxygen, their waters do not. Thus the bottom environment at the soil/water connection teems with microbes and other micro-decomposers which separate, breakdown and neutralize pollutants. Nitrogen is converted to gas and lost to the air; heavy metals are taken up by plant roots for storage in plant tissue or are locked up in soils. Oils and other toxic chemicals are held until chemically decomposed.

While uplands may change rapidly due to plant succession, grazing, and species introduction, wetlands remain constant over long periods of time. This durability is crucial to water quality and the many creatures who lend color and meaning to our world.

Additionally, wetlands maintain open space in urbanized areas, providing wildlife habitat and passive human recreational opportunities while enhancing nearby property values.

Grit chambers are expensive, handle only heavy sediments and must be maintained at considerable cost. Material from a grit chamber is technically hazardous waste and cannot be dumped back into the lake.

Constructed and maintained properly, there is no substitute for nature's way in the defense of water quality. But can a wetland be rebuilt to replace those imprudently removed?

Within well-defined limits, the answer is 'yes.' Constructed wetlands can't perform all the miracles of natural wetlands (the cost of wetland restoration incidentally, is much cheaper than the cost of wetland creation), nor can they do them as well, but for purposes of stormwater treatment, they are without a doubt, the best alternative. In fact, the closer engineers can duplicate natural wetland function, the greater the stormwater treatment that takes place. The major drawback to wetland reconstruction, is the relatively large area required.

Wetland planning first requires a determination of how many acres of upland drainage is to be served and what average volume of stormwater runoff can be expected? This is important since for proper pollutant removal the wetland pool must be able to hold all the runoff from a 1.5 inch storm (the 'average' rain event is .5") at least 24 hours.

Recommended surface area for constructed wetland stormwater basins should be 1-3% of the watershed area. Thus a drainage area (watershed) of 1,000 acres would require 10-30 **total** wetland acres for maximum treatment effectiveness. But small wetlands add up to big results and can be grown like pearls on virtually any amount of existing shoreland, producing a necklace of scrubbers throughout the course of this invaluable community resource.

Though treatment effectiveness would vary, problems of wetland reconstruction are generally political rather than structural.

Good wetland design consists of a two-stage arrangement: 1) a *forebay*, or deep pond for extended wet detention of stormwater; and 2) *vegetative cells* to 'polish' the treated stormwater.

The *forebay* is a pool of permanent

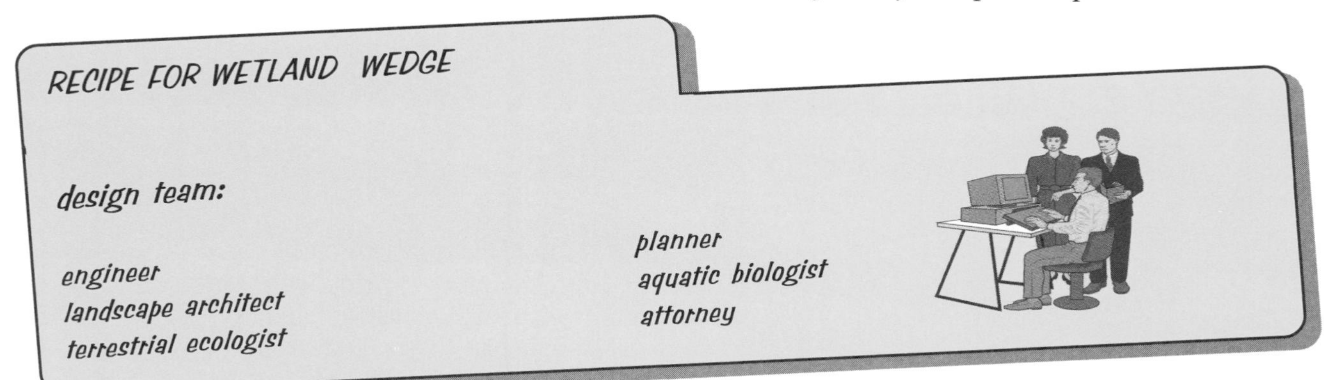

Man-made wetland and a thing of beauty.

standing water 3-6' deep where stormwater enters, flow energy is reduced and larger sediment particles settle out. The pool is limited to 25% of the wetland's water volume in order to maintain oxygen so phosphorus is trapped in sediments rather than released.

The *forebay* has two other necessary features: a) a hard bottom for regular cleaning; and b) a control structure to separate it from the rest of the wetland to prevent immediate pass-through of pollutants in all but the heaviest rainfalls. Filtering in the forebay reduces the pollutant load to tiny pieces of leaf litter.

As stormwater enters the forebay it pushes standing water into the second stage. This is a series of shallow *vegetated* holding *cells*, less than 1' in depth, covering about 3/4s of the total wetland acreage. They're shallow to accommodate lightly rooted vegetation which will absorb nutrients before the treated product reaches the final outlet into the main lake.

The first cell, or 'low' marsh, may be planted with pickerelweed, arrow arum, water hyacinths, and wild rice - plants whose minute hairs trap small pollutants suspended in the flow. Other cells might be 'high' marsh vegetation: lizards tail, sweet flag cutgrass, cattails and reedy bulrushes - plants which absorb large amounts of nutrients nitrogen and phophorus through vigorous growth within the wetland.

Stormwater which reaches the lake after a week of such grooming is similar to the secondary treatment outcomes accorded most sanitary sewage - often cleaner. (Unfortunately, many rare wetland types - fens, bogs and sedge meadows - are being plowed under for development while common types - cattail marshes - are being created to deal with the loss.)

The challenge is to make wetlands be 'natural' with native plantings, gradual rather than steep slopes, and an aversion to long, straight lines. An experienced contractor is essential to implement specific instructions regarding grading, slopes, erosion control, seed bed preparation, seed mixes, and construction sequencing.

As construction begins, plant ecologists select the plant types necessary for the creation of a wildlife community in the wetland. Upland soils are then excavated and replaced with natural organic wetland soils for the saturated conditions required by the *hydrophytic* (water loving) plant community. (To operate wetlands during freezing weather, special care must be taken to direct inflow under the ice and to keep the pond's standing water level high enough to allow wetland function below the ice. If spring snowmelt is allowed to run on top of the iced-over pond (ice is often 2' thick), nutrient removal may fall well short of the 60-80% typical of wetland detention.)

A wetland plant perimeter some 20' wide is established surrounding the wetland. This apron is temporarily flooded during rain events. On its outer edge, it is planted with high grasses which act as a natural fence to discourage grazing geese and wading children. It is sloped on the inside edge to eliminate the tiny stagnant puddles preferred for mosquito breeding. In addition to gradient controls, larva-eating fish can be established in the wetland as a further check on this miniature molester.

Another pestilence - carp - can be captured and permanently 're-educated' by means of a fish weir at the wetland's lakeside outlet. Using the carp's spawning instinct to seek shallows in spring, a trap can be set that will slow its reproduction and the damage it can do. (Minneapolis' first attempt at reconstructing a wetland was destroyed when several hundred carp were allowed to invade and uproot newly set vegetation.)

A landscape architect integrates the treatment aspect of the project into an aesthetically pleasing whole by adding upland contours, compatible shrubs and flowers, bicycle paths, bridges, viewing points, picnicing and play areas. Over the last few decades homes near wetland ponds have proven to be worth as much as 28% more than real estate without this amenity.

Attorneys ensure that all local laws and regulations are met and questions of public safety and liability addressed.

And there are the citizens who, despite some initial apprehension, are willing to hold lake health pre-eminent. Their involvement spreads understanding of project objectives among the public, and helps to refine wetland design to alleviate possible complaints.

If there is not enough room (including using a part of existing open water) to recreate wetland values, there are dozens of structural Best Management Practices (BMPs) that can be employed to scrub runoff but individually they fall short of complete wetland treatment. Not only do BMPs other than wetlands have less impact over a smaller range of pollutants, they are often more expensive, demanding strict maintenance intervals.

Possible watershed modifications can be structural (depending on space available) or cultural, i.e., re-shaping practice. A short list includes: dry basins (storage less than 6 hours is of limited value to water quality); extended detention dry basins (moderate particle storage, less reliable nutrient removal); grit chambers (very little removal of trace metals, fine sediment and nutrients); street sweeping (most effective for removal of coarse particles, leaves and trash), brush sweepers may actually worsen small particle loads to lake! Street sweeping removes about 17% of phosphorus, nitrogen and solids at a cost of roughly $600 per lb. vs. efficiencies of 50% or better at a cost of $400 per lb. for wetlands.)

After installation, wetland costs decrease over time, while outlays for many other BMPs increase. Also, these lesser options need to be used in multiple combinations to approach the effectiveness of wetlands.

COMING TO TERMS WITH STORMWATER

Aquatic - Found in water; related to water.

Bioaccumulation - Build-up of toxic substances in fish flesh. Toxic effects may be passed on to humans eating the fish.

Biomanipulation - Adjusting the fish species composition in a lake as a restoration technique.

Carbon - Abundant natural element in organic (living or once-living) matter.

Compost - A mixture of decaying organic matter useful as fertilizer.

Ecosystem - The interaction of all living things - animal, plant and microorganisms - with each other and with the surrounding environment.

Erosion - The wearing away of soil and rocks by water or wind.

Eutrophication - Increases in mineral and organic nutrients reducing dissolved oxygen, producing an environment which favors plant over animal life.

Fertilizer - Nitrogen, phosphorus, potassium and other compounds spread on or worked into soil to increase its fertility.

Limnetic Zone - A lake's open water habitat for phytoplankton, zooplankton and fish.

Littoral Zone - A lake's fertile shallow shoreline waters dominated by green plants, table and nursery for much of aquatic life.

Macroinvertebrate - An animal visible to the unaided eye but lacking a backbone.

Mesotrophic - Middle fertility; between eutrophic and oligotrophic.

Microorganism - An animal or plant of microscopic size (example: bacteria).

Mulch - A protective covering placed around plants to prevent evaporation, freezing and competition from weeds.

Nitrogen - A nonmetallic element comprising 4/5 of the air by volume; assists plant growth and is commonly used in fertilizers.

Nonpoint Source Pollution - Pollutants that do not have a single source. Nonpoint source pollutants originate from sidewalks, farms, lawns, streets, parking lots, construction sites, and many other centers of human activity.

Noxious - Injurious to health.

Oligitrophic - Low-nutrient, cold, deep and clear basins with high dissolved-oxygen.

Organic - Composed of animal or vegetable

matter.

Pesticide - Any herbicide, insecticide, fungicide or rodenticide which kills without thinking.

Phosphorus - A highly reactive, poisonous, nonmetallic element occuring naturally and used in a wide variety of commercial products; a key ingredient in plant growth.

Phytoplankton - Algae, the base of the food chain, increases lake oxygen when living, depletes it when dead.

Photosynthesis - The process by which green plants produce oxygen from sunlight, water and carbon dioxide.

Pollutant - Any gaseous, chemical or organic waste that renders noxious air, soil or water.

Runoff - Rainfall or snowmelt that is not absorbed into soil but runs high and dirty into your lake.

Sedimentation - Soil and mineral particles (sand) entering a lake in large quantities, smothering fish eggs and making the basin shallower and warmer.

Storm Drain - A hole in the street where stormwater ditches the surface realm for a swift slide into lake or stream.

Trophic Status - The level of growth or productivity of a lake as measured by phosphourus content, algae abundance, depth of light penetration and upset swimmers.

Watershed - That area of land that drains into a specific body of water such as a stream, lake or ocean.

Wetland - An area of land that is covered by water for at least part of the year supporting a wide variety of plants adapted to wet feet.

Zooplankton - Microscopic animals.

A SPECIAL THANKS TO OUR SOURCES

Center for Marine Conservation • Metropolitan Council Environmental Services • Minneapolis Park and Recreation Board • Minnesota Pollution Control Agency • University of Minnesota Limnological Center • Freshwater Foundation • Minnesota Dept. of Agriculture • Minnesota Department of Natural Resources • Friends of Environmental Education Society of Alberta • Izaak Walton League of America • National Geographic Society • Minnesota Department of Health • Minneapolis Public Library • Hennepin County Libraries • Ringer Corporation • Minneapolis Solid Waste and Recycling • Minneapolis Water Quality Committee • Minnehaha Creek Watershed District • Sustainable Resources Center • Terrene Institute • University of Wisconsin-Extension (UWEX) • Wisconsin Department of Natural Resources • U.S. Environmental Protection Agency (EPA), Assessment and Watershed Protection Division, Office of Wetlands, Oceans and Water • United States Department of Agriculture (USDA), Soil Conservation Service • Minnesota Public Lobby

BOOK'UM!

COMPOSTING

Let It Rot! (Stu Campbell) **Organic Gardener's Composting** (Steve Solomon) **The Rodale Book of Composting** (Rodale Press)

IN-HOME

Clean & Green, The Complete Guide to Nontoxic and Environmentally Safe Housekeeping (Annie Berthold-Bond) **Baking Soda Bonanza** (Peter A. Ciullo) **57 Ways To Protect Your Home Environment (and Yourself)** (Univ. of Illinois) **Clean House, Clean Planet** (Karen Logan) **Dan's Practical Guide to Least Toxic Home Pest Control** (Dan Stein)

MULCHING

The Mulch Book (Stu Campbell)

LAWNS

Building A Healthy Lawn (Stuart Franklin) **Natural Lawn Care** (Dick Raymond) **The Lawn, A History of an American Obsession** (Virginia Scott Jenkins) **The Chemical Free Lawn & Garden** (Warren Schultz) **Smart Yard 60-Minute Lawn Care** (Jeff & Liz Ball)

NON-LAWNS

Waterscaping (Judy Glattstien) **Stonescaping** (Jan Kowalczewski Whitner) **Landscaping For Wildlife** (Carrol Henderson) **Creating A Butterfly Garden** (Marcus Schneck) **Butterfly Gardening, Creating Summer Magic In Your Garden** (The Xerces Society/Smithsonian Institution) **Noah's Garden, Restoring Back Yard Ecology** (Sara Stein) **Redesigning The American Lawn** (F. Herbert Bormann, Gordon T. Geballe, Diana Balmori) **The Wild Lawn Handbook, Alternatives to the Traditional Front Lawn** (Stevie Daniels) **Rock Gardens** (The New York Botanical Garden) **Evening Gardens** (Cathy Wilkenson Barash)

Continued on page 116

Continued from page 115

PEST CONTROL

The Gardener's Bug Book (Barbara Pleasant) **Bugs, Slugs & Other Thugs** (Rhonda Hart) **Nature's Outcasts** (Des Kennedy) **Common-Sense Pest Control** (William Olkowski, Sheila Daar & Helga Olkowski) **The Organic Gardener's Handbook Of Natural Insect and Disease Control** (Ellis and Bradley) **Introduction to Insect Pest Management** (Robert L. Metcalf and William H. Luckmann) **Good Bugs for Your Garden** (Allison Mia Starcher) **Carrots Love Tomatoes: Secrets of Companion Planting** (Louise Riotte) **Pest Control You Can Live With** (Debra Graff)